Instructor's Resou...

David T. Marx
Southern Illinois University at Carbondale

Physics

Sixth Edition

John D. Cutnell
Kenneth W. Johnson
Southern Illinois University at Carbondale

WILEY

John Wiley & Sons, Inc.

To order books or for customer service call 1-800-CALL-WILEY (225-5945).

Copyright © 2004 by John Wiley & Sons, Inc.

Excerpts from this work may be reproduced by instructors for distribution on a not-for-profit basis for testing or instructional purposes only to students enrolled in courses for which the textbook has been adopted. *Any other reproduction or translation of this work beyond that permitted by Sections 107 or 108 of the 1976 United States Copyright Act without the permission of the copyright owner is unlawful. Requests for permission or further information should be addressed to the Permissions Department, John Wiley & Sons, Inc., 111 River Street, Hoboken, NJ 07030.*

ISBN 0-471-23128-2

Printed in the United States of America

10 9 8 7 6 5 4 3 2 1

Printed and bound by Victor Graphics, Inc.

PREFACE

Resources: More than an Instructor's Guide

If this is your first time using Cutnell & Johnson's *Physics*, welcome. A quick glance of the table of contents reveals that it has been organized in the now standard sequence, but a closer look within reveals the features that have helped make this the leading college physics text. The book is enjoyed by students for its readability, smart design, and thorough explanation of concepts and problem-solving strategies. Features that instructors appreciate are the careful development of physical principals, reasoning strategies, conceptual examples and questions, and thought-provoking homework problems with a range of difficulty. Students will appreciate the vast array of supplementary material available on the internet from Wiley at www.wiley.com/college/cutnell.

The things that I felt instructors, beginners or veterans, would want in an instructor's guide are:

▷ tips for general course management (syllabi, policies, grading, exams, etc.)
▷ a comprehensive guide to the internet and its physics resources
▷ advice on using lecture demonstrations, computers, and audio/visual technologies
▷ teaching strategies
▷ tips for getting the most out of the instructor's supplementary text materials

Other features that have been retained include: lecture planning and notes, a textbook conversion table that compares the leading algebra-based physics texts, a list of available transparencies, a homework problem map from the fifth to the sixth editions of the text, films, lecture demonstrations, laboratory exercises, computer simulation programs, and suggested usage of the text.

A discussion of physics teaching strategies can quickly turn into a heated debate. Every instructor develops his/her own teaching method or style from experience as both student and educator as well as their own uniqueness as human beings. Some people are more attuned to the modes in which information is communicated and learned, make careful mental notes of what works or doesn't, and continually seek better methods. Others enter into teaching without ever having thought much about it and simply review the text in a simple lecture format with little additional effort. As we all know, the spectrum is broad indeed. The approaches and advice given in this book are naturally affected by my own biases. In preparing this manuscript, information has been drawn from a number of sources including the instructors who taught me, past and present colleagues, e-mail discussion groups, selected physics education literature, and resources available on the internet.

Dedication

This book is dedicated to the exceptional people who move us to new levels of understanding of both ourselves and the universe we live in and ask us to find the very best in ourselves.

Acknowledgments

My initial thanks go Robert Lee Kernell, author of the third edition of this book, and to J. Richard Christman, author of the *Instructor's Manual* for *Fundamentals of Physics, 5e.* by Halliday, Resnick and Walker. I found both books to be a valuable starting point in the creation of this book in its fourth edition.

I am indebted to John D. Cutnell and Kenneth W. Johnson for their continued support, ideas, and inspiration. Many thanks also go out to the many internet colleagues mentioned above who share in the joy of teaching and in the love of physics; your sharing of knowledge, sometimes in heated discussion, gave me many insights (and, occasionally, laughs). As always, it has been a pleasure to work with all of the people at John Wiley & Sons, but special thanks go to Stuart Johnson and Geraldine Osnato.

David T. Marx

Southern Illinois University at Carbondale
Carbondale, Illinois 62901
e-mail: siucafs@netscape.net

Contents

1 Getting Started: Course Organization — 1

- Course Planning and Policies — 1
  - Course Goals — 2
  - Students and Their Needs — 3
  - Course Content — 4
  - Teaching Assistants — 5
  - Quizzes, Tests, and Exams — 5
  - Homework Assignments — 8
  - Grades and Curves — 8
  - Attendance Policies — 9
  - The Syllabus — 10
  - Cheating and Its Prevention — 11
- The Lecture — 14

2 Physics Resources on the Internet — 17

- The World Wide Web (www) — 17
- Search Engines and Portals — 19
- E-mail and E-Mail Discussion Groups — 20
- Physics-related Resources on the World Wide Web — 23
  - Organizations, Newsletters, & Journals — 24
  - Physics Education — 25
  - Web-based Courses & Lecture Notes — 26
  - Lecture Demonstrations — 27
  - Audio/Visual & Computer Simulations — 27
  - Miscellaneous — 29

3 Print, Audio/Visual, and Computer Resources — 31

- Print Resources — 32
  - General Pedagogy — 32
  - Laboratory — 34

	Computers in Physics	35
	Mechanics	37
	Thermodynamics	45
	Wave Motion	47
	Electricity & Magnetism	48
	Light & Optics	51
	Modern Physics	54
Audio/Visual Resources		56
Computer Software		64

4 Usage of the Text and Supplementary Materials   67

The Semester System	67
The Quarter System	69
Getting the Most from the Text Supplements	73
Student Study Guide	73
Student Solutions Manual	74
Instructor's Solutions Manual	74
Test Bank & Computerized Test Bank	75
Homework Disk	75
Transparency Acetates / *Take Note!*	76
Instructor's Resource CD	76
Cutnell & Johnson Website	76

5 Lecture Planning & Notes   79

1	Introduction and Mathematical Concepts	81
2	Kinematics in One Dimension	84
3	Kinematics in Two Dimensions	87
4	Forces and Newton's Laws of Motion	90
5	Dynamics of Uniform Circular Motion	94
6	Work and Energy	97
7	Impulse and Momentum	101
8	Rotational Kinematics	104
9	Rotational Dynamics	108
10	Simple Harmonic Motion and Elasticity	111
11	Fluids	115
12	Temperature and Heat	119
13	The Transfer of Heat	123
14	The Ideal Gas Law and Kinetic Theory	126

15	Thermodynamics	128
16	Waves and Sound	133
17	The Principle of Linear Superposition and Interference Phenomena	137
18	Electric Forces and Fields	141
19	Electric Potential Energy and the Electric Potential	145
20	Electric Circuits	149
21	Magnetic Forces and Magnetic Fields	154
22	Electromagnetic Induction	159
23	Alternating Current Circuits	163
24	Electromagnetic Waves	166
25	The Reflection of Light: Mirrors	169
26	The Refraction of Light: Lenses and Optical Instruments	173
27	Interference and the Wave Nature of Light	178
28	Special Relativity	182
29	Particles and Waves	185
30	The Nature of the Atom	190
31	Nuclear Physics and Radioactivity	194
32	Ionizing Radiation, Nuclear Energy, and Elementary Particles	198

Appendix A

College Physics Textbook Comparison and Conversion 202

Appendix B

Homework Problem Locator 206

CHAPTER 1

Getting Started: Course Organization

This chapter is meant to help both experienced and beginning instructors with various aspects of course organization and presentation. The chosen topics are wide-ranging, but not comprehensive. Much of the discussion naturally comes from my own experience both as a student and as an educator. The information also comes from discussions I've had with other educators and students as well as from various e-mail discussion groups. Therefore, because of the subjective nature of the material, there is always room for discussion about the best way to handle a particular course policy or teaching approach. Alternative approaches are suggested, wherever it is possible. I hope you'll find value in the following pages and information that will make your course run more smoothly. As educators, we all try various methods to instruct effectively. Some of these work; and some do not. New approaches, especially ones that move away from the traditional lecture format, tend to be more dependent on the individual instructor for success than on the techniques themselves. My only advice when it comes to choosing an approach to teaching is this: *always seek out ways to improve your teaching, find out what works for you and your students and discard the rest, and try not to lose your enthusiasm along the way.* Chapter 2 of this book lists internet resources for physics educators including web sites and e-mail discussion groups; and chapter 3 provides an extensive bibliography on pedagogy.

Course Planning and Policies

The semester is starting next week and you are just starting to think about the algebra-based introductory physics course you've been assigned to teach. Perhaps it's one of several courses you'll be teaching. If you've taught the course before, perhaps you've thought of some new things you'd like to incorporate or you are looking for ways to avoid some of the problems you have had in the past. Or, if you are new to teaching, this section will help point out some of the pitfalls of teaching. In either case, one quickly realizes that there's more to teaching than just preparing and delivering lectures. This section will point out these pitfalls and suggest ways of avoiding them. There is no substitute for good course planning; and this time is well spent before the course gets underway because it will help minimize policy-related problems that can interfere with both teaching and learning.

Here are some questions you may want to consider in starting your course planning:
1. What are my overall goals for the students in the class?
2. What are the backgrounds of the students? What are their needs?
3. What chapters do I want to cover? Which sections can be omitted? Which is more important, depth or breadth?
4. Are there any new teaching strategies that I want to try?
5. How do I want to utilize graduate assistants? (*if applicable to your institution*)
6. How many exams (quizzes) should there be?
7. Will the assigned homework problems be graded? Should I use multiple choice answer sheets or grade the content of each problem?
8. How will the overall grade be determined? What will the initial grading scale be and should a grading curve be used?
9. What are some ways that I can maintain a high rate of attendance throughout the semester?
10. How can I use lecture demonstrations, computer simulations, and audio/visual resources to improve the course? To what extent do I want to use the internet?
11. How will I handle cheating and prevent it from occurring?
12. What items should students buy other than the textbook? Do I want to specify the type of calculator students may use for exams?
13. Are there any materials I want to put on reserve at the library?

Course Goals

The algebra-based physics course is one opportunity we have as physicists to reach out to non-scientists and to prove the value that physics can have for everyone. It is about public relations as much as it is about education. Our goal is not to turn everyone into a physicist, but rather to start individuals on the road to critical thinking, to better problem-solving ability, and to become better citizens in this increasingly technical world. Long after students leave our courses, they may not remember all the phenomena and equations, but they will remember whether or not they enjoyed the course and, perhaps, they will recognize some of the tools that they carry with them and use.

Here are some goals to consider:
- to dispel the popular notion that physics is only accessible to smart people and to show that physics is interesting, useful, and valuable (physics is fun and cool) to everyone
- to dispel the popular notion that physics is just a collection of equations used to solve obscure problems
- to challenge students to improve their abilities to reason, to think critically, and to solve problems
- to help students gain conceptual understanding of the physical universe by studying the interrelationships between the concepts that are the framework of physics and to apply mathematical tools as an aid to understanding those relationships

Students and Their Needs

Students come into the physics classroom with a different backgrounds, levels of ability, preconceptions, reasons for taking the course, and needs. In my algebra-based physics courses, I've had students from a spectrum of majors including: biology, pre-medical, pre-veterinary, computer science, philosophy, administration of justice, engineering technology, humanities, continuing education, pre-law, and the undecided. Students have ranged from seventeen to sixty-three years old and have represented nations from all continents. With all of this diversity, one cannot possibly account for or address all of the varied life experiences of students, but there is always some common ground on which the entire course can be built. This is a necessary first step; and it must occur as early as possible in the course or students will be quickly left behind, feel alienated, and drop out. The easiest way to do this first step is to point out that you recognize the differences in backgrounds, abilities, preconceptions, reasons for enrolling the course, and needs. Point them out and discuss as many of them as you can. The students will know that you understand and are accessible. Tell them your goals and your plan for helping them meet the challenge you've laid out for them.

Respect the individuality of students. Take time to learn their names and something about each of them, whether you have a class of thirty or two hundred. Students will be more responsive, feel less inhibited about asking questions, and make a greater effort to succeed in the course if you show you care about them and their needs.

Most students take the course because they are required to take it as part of the requirements for their major. Point out that there is a definite reason why physics is required for their majors and that they are expected to gain certain skills in this course that they will probably not get anywhere else. Often times, students are under pressure to achieve a particular (high) grade in the course, especially pre-med. or pre-vet students. The problem then becomes moving the students away from their focus on grades to a focus on learning. Unfortunately, in this era of diminishing standards and grade inflation, students are generally not shown a correlation between doing the work to learn and grades.

Set your standards as high as you like, students can be motivated to rise to the level you set (with enough guidance and assistance from you, teaching assistants, and other students). Reassure them that everyone has an equal chance of succeeding in the course (the meaning of *success* is subjective), but not everyone will get an A. The reward for everyone is whatever they are able to carry with them from the class, not the grade. I tell students that they will get through the course, complete their majors, and get their degrees; but then I ask them what will make them stand out from the crowd after graduation when they are trying to get their first jobs or a year after starting that job? The answer isn't their physics grade or their GPA, but rather the skills they carry with them that others didn't bother to do the work to achieve. Sure, students will come to the course with unrealistic expectations ("I can avoid going to lecture by getting notes from someone else, then read the notes five or six times in the two hours before the exam, skim the text, and get an A."); and we can't expect to reach everyone in a class of two hundred with our message, but it is well worth the effort.

I've mentioned the above philosophy to some colleagues and some have asked, how do you deal with students that come from poor educational backgrounds and lack necessary math and/or language skills? My view is that it is *my job* to teach, not just physics, but also anything else that will help the student understand and learn the material better. I give my students as much time in my office as they need. I let them know that I'm always available for their questions about anything at all (not just the next homework assignment or quiz). I've used the pronoun "I" in the previous few sentences because I know this isn't for everyone. Many instructors have too many obligations outside of the classroom, have only specific office hours, and cannot find more time for students. There's nothing wrong with that approach either. At least point the students to a place where their questions can be asked and answered, such as a tutoring center or a graduate assistant's office.

For an instructor, it's easy to forget that students are taking a number of other courses and have responsibilities to each. Students tend to only make an effort toward those things that affect their grades and, generally, this is done at the last possible moment. If you want students to change a behavior, make it part of their grade.

Course Content

Each instructor sets his or her goals for the students in the course; and depending on those goals, the amount of coverage and the order of the topics must be determined. You may have 60 to 75 hours of contact time to accomplish these goals within the academic year. The text contains 32 chapters of material, organized in the somewhat standard sequence: mechanics, thermal physics, waves, electricity and magnetism, light/optics, and modern physics. To cover the entire book, one would use only two to three hours of contact time per chapter. This can be done, but few instructors probably attempt it; and it can only lead to frustration for almost all of the students. [For students preparing for the MCAT, it is important to cover all of the major topics in the book to some extent. I had one student come back after the MCAT and thank me for getting through Chapter 32 because two questions on the exam came from the material that we had just discussed in class two weeks before the MCAT.] Material in the text that may be omitted with little impact to the overall development of the material is either placed in separate sections, denoted with an asterisk, or at the end of a main topic section. Chapter 4 of this book provides suggested schedules for both the quarter and the semester systems.

Some people argue that college physics courses have become a survey of topics without the necessary depth to provide an understanding of what physics is about or the scientific process that allows new knowledge and understanding to be acquired. The ideal then is to find a balance between breadth and depth that meets both the goals of the instructor and the needs of the students. An approach to maintaining a sense of structure to the course without physics appearing to be just an encyclopedic compilation of physical phenomena is to follow the connection between the various topics via the *Concepts at a Glance* feature. Chapter 5 of this book is designed as an aid to your course planning and includes a description of the *Concepts at a Glance* figures in each chapter. All of the sections of each chapter of the text are listed as well as teaching objectives, available transparencies, internet resources, laboratory experiments, and lecture demonstrations.

Teaching Assistants

Teaching assistants (TAs) can be both a benefit and a hindrance to teaching. One thing that we may lose as instructors in using teaching assistants is the ability to directly monitor students' progress on homework assignments, quizzes, and laboratory exercises. We may not see how students interpret problems, how they answer them, nor hear the questions they are asking. How can we gauge our effectiveness in teaching without being in touch with our students?

Teaching assistants can lessen an instructor's burden by doing grading, supervising laboratory sections, handling tutoring sessions, conducting discussion (recitation or quiz) sessions, and proctoring exams. Using TAs can also lead to problems. For example, if more than one person is involved with grading, there will likely be inconsistencies in grading that students will notice. Another problem can arise if the assistants do not attend the lectures. Students in tutoring or discussion sections may be confused by assistants using notation or explanations that differ from those used in class or in the text. Requiring that everyone use the notation and language in the text will minimize confusion.

To further reduce potential problems, meet with the assistant(s) and discuss your approach to teaching, course policies, grading procedures, and your expectations of them. Ask them to attend your lectures. Ask them to read the text before meeting with students to discuss the current material. If they teach laboratory sections, ask them to perform the experiment (individually or as a group) before attempting to lead others through it and to examine each experimental station before class time.

Quizzes, Tests, and Exams

No matter which word you choose *quiz*, *test*, or *exam*, nothing elicits a student's fear response more quickly than having to take one of these. The connotative meaning of these three words varies, but the word *quiz* seems to be the gentlest. There are many ways to test students over the course of a semester. I'll try and mention as many ways as I can think of in the paragraphs that follow, but here is how I conduct testing in my classes: a 15 to 20 minute quiz is given every other week. This results in six or seven quizzes per semester; and lowest quiz score is not counted toward the course grade. The quiz covers one or two chapters and is divided into two parts: conceptual questions (two or three) and problems (two or three). For the conceptual questions, students are asked to write a short answer (no fill-in-the-blanks or true/false-type questions). For the problems, students are asked to show how the problem is set-up (drawing and/or free-body diagram), to show and apply the correct equations for the situation, and to choose the correct units. Students may use their own class notes (no photocopies allowed) for the quiz. [This is one technique I use to maintain a high rate of attendance at lectures.] The final *exam* is comprehensive; and students are allowed to use the text, their homework assignments, and their class notes.

Students may also use a non-programmable, scientific calculator (see the discussion on calculators below) for both the quizzes and the final exam.

Number of Tests

Because of large class sizes, many instructors prefer to give only two or three exams, including the final. Unless these are given in a separate section, one has to give up two lecture sessions to give these exams. Six or seven quizzes require about the same amount of lecture time, but students are under less stress because they believe they have a greater chance at succeeding in the course when there are seven quizzes than they do if they have two mid-term exams. Everyone has a bad day from time to time when we're just not at our best performance for one reason or another. With several quizzes, a student can have one such bad day without seriously affecting his or her chance at proving herself/himself over the course of the semester (especially if the lowest grade isn't counted). It's really just an illusion since the same number of questions are asked and the same number of points are given whether quizzes or exams are given. The difference here is that students are more comfortable, less fearful, and more likely to focus on the content of the test and spend less time dwelling on whether or not they will pass it.

Multiple Choice versus Written Tests

Again, if saving time is a necessity, the choice is clear. Multiple choice (MC) tests are graded by computer and the results are immediately returned with all the statistical information one could ever want. MC test grading, therefore, doesn't have the subjectivity that grading of a written test has (especially if more than one grader would be used). A test bank containing about 2200 multiple choice questions is available from your Wiley representative along with excellent test preparation software as supplements to the Cutnell & Johnson text.

I prefer to use written tests for the following reasons:
- I can get an idea of the student's thought process on conceptual questions since the answers are in her or his own words. Adjustments to the manner in which concepts are introduced in lecture may then be made to clear up recurring misconceptions.
- I can write comments directly to the student suggesting ways to improve their performance. I can even give direct praise to a student that seems to really understand the material well or has come up with an answer that I had not considered, but is nevertheless correct.
- I can see if the student is using a viable problem-solving approach and give partial credit for all correct steps. On a multiple choice test, credit is all or nothing. The scenario often occurs that a student understands the problem, takes a correct approach, but then a small math error occurs (or the wrong button is hit on a calculator), and the wrong answer is obtained. [Some MC test systems can be set up to give partial credit for specific wrong answers.] Sometimes, the student merely shades in the wrong circle.

- In looking over the tests, one can easily see which students are doing well and which students need help (and exactly what kind of help is needed). A short discussion with the student in need of help often further clarifies the problem and leads to a satisfactory resolution. If a "drowning" student goes unnoticed, he or she will likely drop the course.

Open- versus Closed-book Tests

Memorization has never been a cornerstone of physics. Most physicists value conceptual understanding and problem-solving skills above memorization. We do tend, over time and repeated encounters, to remember physical constants, solutions to particular problems, mathematical formulas, equations, the particular wording of a physical law, etc. A student generally doesn't have the necessary time to have repeated encounters with material within the course to achieve that kind of familiarity, so for tests, instructors usually choose a method to help the student with lists of physical constants, formulas, and equations.

Here are some methods (listed in order of increasing openness):
1. A list of equations and physical constants is given to the students along with the test. Sometimes this list includes additional equations or constants that are not applicable to the material on the test.
2. A single sheet of paper (or some fraction) is prepared by the student and may include anything the student thinks will be useful. Instructors sometimes place limits on what may be on the sheet, such as: "no solved problems" or "no photocopies."
3. Students are allowed to use their written class notes. Presumably, these notes contain all necessary constants, equations, and examples presented in lecture. Students are thus encouraged to maintain complete and well-organized notes. Students may also be allowed to use their own homework assignments. Photocopies should not be allowed because it does not encourage the student to read the text, to attend lecture, to think about the material, or to keep a notebook. If a student has written out their notes (even copying portions of the text), the information has at least has passed through his brain (at least once) and onto the paper.
4. Students are allowed to use the entire textbook, but they may not use any written notes or homework assignments. This gives students all available equations (including the mathematical formulas in the appendix) and physical constants. Students must know their book well in order to find the needed information quickly.
5. Students may use the textbook, their notes, and homework assignments. This has all of the advantages listed in numbers 1 through 4 above.

The greater the degree of openness of the test, the more students will rely on that openness and do less preparation for the test. Students who are ill-prepared can be observed taking an open book test, loudly flipping pages and feverishly searching their (entire) text for an equation that looks like it will work or an example problem that resembles the test question. I warn my students time and again about not relying too heavily on the openness, but rather to view their notes as something to fall back on if something is forgotten or as a security blanket. If they have written their own notes, they don't have to search for something; they know exactly where they have written it. Lastly, as I indicated before, using open notebooks encourages attendance (especially if material is given, and later tested, that is not in the text).

Homework Assignments

Homework is the place where students have an opportunity to be challenged and to learn physics. Students, of course, do not realize this (in general); and many seek ways to minimize their effort on homework assignments. The actual effort is proportional to the value the homework has toward the course grade and the seriousness with which the instructor treats the assignments. Therefore, the instructor must determine how much of his or her time will be put into grading the assignments. One time-saving solution is to use the multiple choice homework answers provided in the *Test Bank* and the accompanying software.

Students usually work in pairs or in groups on homework assignments (a cooperative effort sometimes fosters learning better than that within a competitive environment). This should probably be encouraged with the warning that students can help each other, but no one should rely on anyone other than himself or herself to learn the physics. Reinforce the notion that learning and corresponding good grades are the result of doing the necessary work. You'll still find students copying answers from each other up until the very last minute (a form of cheating, to be sure), but some will get the message (sometimes after it is pointed out that their "high" homework average does not result in a "high" quiz average).

Grades and Curves

Grades are much less meaningful and less important than students believe, but one cannot easily change the popular perception. Some educators fantasize about what it might be like to teach in a world without grades, a place where diligent students are "thirsty" for knowledge. Some instructors and even some schools have

> **grade** (grād) *n.* - an unsubstantiated report by a biased and variable judge of the extent to which the student has attained an undefined level of mastery of an indeterminate amount of material.*

been bold enough to attempt teaching without grades, but most of us simply face the day to day inquiries, such as: "What do I need to get an A?" or "What's my current grade?" Students are indeed focused on their grades due to both imagined and very real pressures to obtain a particular grade and to maintain a particular grade point average. The instructor needs to be aware of and understand the student's focus on grades and accommodate the student. The student's motivation, desire to work, and performance is partially based on the student's sense of security, often based on knowing his or her current standing in the course.

---

* This is adapted from the original version by Paul Dressel, *Basic College Quarterly*, Michigan State University (winter issue, 1957).

After the evaluation tools of the course have been selected - homework, quizzes, tests, laboratory exercises, reports, final exam, attendance, etc. - one then has to determine the value of each of these. This is very subjective as we all tend to value each of these differently. The rule here is, once again, that if you want students to do something, make that item a substantial part of their overall grade. Once decided, the percentages should be indicated on the course syllabus.

The grading scale is always a subject of contention between the instructor and the students. Some instructors select a grading scale such as (90.0 - 100 %) = A, (80.0 - 89.0 %) = B, etc. and do not alter it in any. Others provide students with a tentative scale and alter it as the course progresses. Sometimes, the scale is revised so that it matches a normalized distribution, or bell curve. The more variable the grading scale is, the less comfortable students feel in the course. They lack the security of knowing their grade at any instant in time as discussed above. My preference is to set a reasonable, but tentative grading scale at the beginning of the semester and tell the students that the ranges may be adjusted by one or two points before the 3/4 point in the semester. The scale is then finalized at that time. If one prefers to scale the grades, a useful method is given by D. H. White, *Am. J. Phys.* **26**, 643 (1958).

I have never had to scale grades in my introductory physics courses, but I have used one for an astronomy course. In that case, I specified on the syllabus that the curve could only be *earned* through attendance, which was checked at random (at any time during the lecture period). In the same course, I also informed students that a percentage (5 - 10 %) of the quiz questions would be related to material only available in lecture. Lecture attendance was greatly increased using these techniques.

Attendance Policies

As indicated in the paragraph above, an instructor must sometimes be creative in finding ways to maintain a high rate of lecture attendance. Many instructors have the opinion that attendance is the sole responsibility of the student; and as adults, they have the choice to attend or not to attend. Others (including myself) have the opinion that in order for an instructor to do his or her job, the student must be present. The student has the responsibility to attend and the instructor has the responsibility to provide incentives for the student to attend. Students will not want to attend a lecture that is simply a monologue summary of the text. These incentives include:

- *lecture time that is engaging, informative, and interesting* - This is very much dependent on the creativity and personality of the instructor. A great deal of the field of physics education research has been to find alternatives to the traditional lecture format. Chapters 2 and 3 provide resources and references for incorporating new elements into the lecture format.
- *lecture demonstrations (or audio/visual media or computer simulations) that are educational, relevant, and entertaining* - Many departments have personnel that develop, maintain, and, sometimes, present lecture demonstrations. These people are a tremendous resource and can help an instructor choose and schedule

demonstrations throughout the semester (or quarter). If you do not have this type of person in your department, resources and references for demonstrations are also listed in Chapters 2 and 3.
- *a safe environment for asking and answering questions* - Students demand and deserve respect. They need to feel that every question that is asked is important and taken seriously by the instructor. Getting a class to open up can be difficult because of students' past experiences. It's much safer to remain silent than to face the possibility of ridicule. If the class will not interact with the instructor or with other students, lecture time will be spent in monologue. One technique that instructors use is to ask a question and simply wait for the first brave soul to hazard a guess at the answer. Silence can be uncomfortable. Depending on the instructor's response to that brave person, the tone will be set for the rest of the course.
- *grade-related measures* - Any grade incentive that is fair will, in general, work.

The Syllabus

The syllabus is a written document distributed to the class at the beginning of the course. It may list general information about the instructor (name, office, phone, etc.), office hours, course policies, the grading scale, homework assignments and due dates, quiz dates, etc. The syllabus is, in essence, a legal contract between the instructor and the students enrolled in the course that the student accepts by remaining in the course. The sense that the syllabus is a contract is becoming more prevalent today than ever before. Departments retain copies of the course syllabus on file in the case that a dispute occurs in the future. Some universities require legal counsel to review the syllabi before they are distributed to the class. Instructors may also require students to sign a copy of the syllabus that is then kept on file.

As is the case with other types of contracts, the participants (i.e. students) will try to find loopholes in the policies and continually ask for renegotiation or special treatment. The instructor's contract is with the entire class, not with individual students. Any changes in an announced policy, therefore, should be carefully considered. Usually, one may simply ask oneself, "Will it be fair to the class if this change is made?" If the answer is yes, then it's unlikely that anyone will complain about the change in policy.

When the syllabus is written, it's a good idea to indicate those policies or due dates that are tentative or flexible. For example, one exam policy might be, "No make up exams will be given unless special arrangements are made before the scheduled exam and the absence is unavoidable." The policy is fair because it is available to everyone. Flexibility lies in the fact that the instructor decides whether or not the absence is necessary and agrees to the special arrangement. There is rigidity in that the arrangement must be made before the exam. The instructor should then be prepared to face the student who misses the exam for an acceptable, but unavoidable reason occurring shortly before the exam.

I indicated above that the course policies may be challenged by students, so it's important to have support from the department and the administration for your policies. When considering new course policies, it's a good idea to find out what other instructors' policies are and discuss the pros and cons of them. Discuss the policies with administrators and make use of the school's legal counsel. While some of these steps may seem excessive, it's worth the effort to minimize the potential for disputes that can be carried into a civil court.

Course policies are meant to provide order and fairness in the students' learning environment. Every policy an instructor chooses has a rationale behind it. It's very important to communicate that rationale to the students and discuss each policy when the syllabus is distributed. Give students the sense that the policies are in their best interest and that *fairness* is the metric used in carrying out policies and in adjusting them. When people understand the rules of the game, they are more likely to play it.

On the next two pages, I've included an example syllabus. The due dates and assigned material are designed to work through all 32 chapters of the text during an academic year consisting of two semesters. The listed homework questions and problems were chosen randomly for this example. Again, this is only an example, your syllabus will be based on your own course organization and policies as discussed above.

Cheating and Its Prevention

You may have noticed specific reference to cheating on quizzes and the final exam on the example syllabus. Cheating occurs despite our best efforts to prevent it. Most often, students cheat because they are overwhelmed and they panic. Some universities have specific policies regarding cheating and it's a good idea to be aware of them and make sure the students understand them. The reputation of the school is dependent on maintaining a high standard of intellectual honesty. Some schools have and enforce an honor code. For a clear listing of cheating and plagiarism instances and policies, see the Carnegie Mellon University policy on the internet at:

http://www.cmu.edu/policies/documents/Cheating.html

If cheating is proven, the punishment is often failure of the course, suspension or expulsion. Proof is sometimes difficult, however; and many instructors avoid pursuing cheaters because of the possible ramifications of being wrong. Cheating can be confronted (and should be) as long as the instructor is discreet. In the few instances that I've discovered cheating, I received a full and sincere apology from the student and he/she accepted the consequences with no argument.

Cheating during a test may occur in a number of ways, but all of them cannot be addressed here. An obvious example would be one student copying information from another student's test (with or without the student's knowledge). If the test is multiple choice, it is difficult to prove cheating occurred, even if all of the answers are identical. The best way to prevent cheating on multiple choice tests is to have multiple versions with no easy way for a student to determine which version a neighbor has. The questions on all versions may be the same, but both the questions and the choices reordered.

Example Spring Semester Syllabus

Physics 202   Spring 2003

Lecture Section I   Tuesdays and Thursdays: 9:35 - 10:50 AM
Lecture Section II   Mondays, Wednesdays, and Fridays: 12:00 - 12:50 PM

Instructor: Dr. David T. Marx
Office and Office Hours: Physics 110
                              Mondays and Thursdays: 3:00 - 5:00 PM
                                   other hours by appointment or drop-in

Text: *Physics (Fifth Edition)* by John D. Cutnell and Kenneth W. Johnson

Grading: Homework   25%
            Quizzes   50 %
            Final Exam   25 %

Grading Scale (*subject to change*)
    A   89 - 100 %
    B   77 - 88
    C   65 - 76
    D   53 - 64
    F   < 53

Rules:

Homework Assignments: Ten problems are assigned per week, of which 6 of the 10 will be graded. Each graded problem is worth ten points. Partial credit will be given whenever possible. Work that is illegible will not be graded. There are 15 assignments; and the lowest grade will be dropped. Homework can be handed in during class or be placed into my mailbox before 4 PM on the due date. Late homework, in general, will not be accepted since the solutions are immediately posted.

Quizzes: The quizzes will cover material presented in lecture and encountered in homework assignments. The quizzes will be open notes, but closed book. You may use a non-programmable scientific calculator. Photocopies may not be included in notes. Seven quizzes will be given, of which, the lowest grade will be dropped. Each quiz is worth 20 points. Anyone found cheating on a quiz will receive a zero for that quiz which cannot be dropped. In general, no make up quizzes will be given.

Final Exam: The final exam will cover all material presented in the course. It will be open notes and open book. A scientific calculator may also be used. Partial credit will be given when possible. Cheating will result in a zero grade for the exam.

Example Spring Semester Syllabus (continued)

Section II

Homework Assignments

Chapter	Conceptual Questions	Problems	Due Date
16	5, 8, 15	6, 24, 27, 45, 54, 61	Jan. 28
18	5, 11, 13	2, 9, 24, 37, 43	Feb. 4
19	3, 10, 11	2, 11, 17, 29, 36, 49	Feb. 11
20	1, 2, 11, 15	4, 9, 20, 37, 48	Feb. 18
21	2, 5, 11	1, 9, 15, 24, 44	Feb. 25
22	3, 4, 10	1, 12, 27, 32, 41, 49	Mar. 10
24	3, 4, 8	1, 10, 13, 18, 38	Mar. 24
25	1, 5, 7	1, 5, 11, 16, 32	Mar. 31
26	4, 7, 19, 32	5, 10, 23, 29, 36, 54	Apr. 7
27	3, 5, 15	3, 9, 18, 24, 32	Apr. 14
28	3, 4, 7	1, 8, 15, 22, 27	Apr. 21
29	1, 7, 8	1, 14, 20, 27, 36	Apr. 28
30	4, 7, 10	3, 9, 20, 29, 38	May 5
31	3, 11	7, 12, 18, 27, 30	May 12
32	1, 5, 8	4, 6, 16, 27, 33	May 12

Quiz Dates

January 30
February 14 and 28
March 13 and 27
April 10 and 24

Final Exam

Section II   Wednesday, May 14, 3:10 – 5:10 PM
Plan to arrive at the lecture hall at 3:00 PM.

If the test is written, detection during grading is very easy if there is a single grader and she is attentive. Students will usually copy a neighbor's paper exactly, with each equation and drawing in the same location on the test. Mistakes are copied as well. How does one tell which student cheated? It's usually easy. The cheater has more erasures, incomplete drawings, numerous sign and math errors (but, miraculously, the correct answer), etc. Often, exponents on the offending student's test are different because they're too small to read at a distance.

If tests are "closed book" and "closed notes," cheating may be accomplished using advanced calculators. An individual may be capable of programming equations, class notes, example problems, etc. into a calculator for use on the exam. This would clearly be unfair to the rest of the class that is abiding by the closed test rule. Some calculators even have infrared transmitters and receivers built in that allow communication between calculators, even across the classroom. This could allow students to take the test cooperatively, again unfair to the rest of the class. One way to prevent this kind of cheating is to insist on the use of a standard scientific, non-programmable calculator. Another way is to use problems that only require algebraic manipulation and disallow the use of calculators (some calculators can also do symbolic math).

Another attempt at cheating I have encountered: a student asks for additional credit for an assignment or quiz that has already been graded and returned. Upon closer examination of the paper, one can see that the student has altered either what is written or has added information. This is combated by one instructor by having the course grader draw a line around each item on the paper and then put an "X" through it.

Getting Started... The Lecture

The lecture is traditionally a rhetorical discourse between an instructor and a somewhat *passive* class. Most classes are conducted in this manner beginning in grade school and continuing through the graduate level. The student is obliged to listen and take notes. The instructor's job in this environment is to lead the class through the material, providing examples and demonstrations that aid understanding. The instructor also *activates* the students by assigning work to be done outside the lecture. At some point, the instructor will evaluate the student's level of understanding and a grade will be issued. We are all familiar with this process and many are fairly comfortable with it. Some physics educators have found the traditional lecture to be ineffective in teaching physics and have sought numerous other approaches. In this section, a modified traditional lecture format will be discussed, but resources for finding other approaches and new ideas are listed in Chapters 2 and 3.

The First Lecture

For some, standing in front of two hundred people and speaking for fifty minutes is a frightening experience. For others, it makes no difference if it is a conversation with one person or a speech given to eight million people on television. We find ourselves somewhere on that spectrum. Becoming an effective educator requires facing any fears we might have and breaking down those barriers that prevent us from communicating fully. Effective teaching and learning requires bi-directional communication (with all parties being good listeners as well). The relationship between the instructor and the class begins with the first lecture meeting. Students attending that first lecture make decisions about what their own level of participation will be in that relationship throughout the course. Some will choose to work very hard in the course; and some may choose to withdraw from it.

An instructor should do some things during the first lecture:
1. introduce herself/himself
2. discuss in general what physics is and, perhaps, what it isn't
3. lay out course goals and indicate why they are worth achieving
4. provide a syllabus, discuss all course policies, indicate policies that are flexible, and note any due dates or grading scales that are tentative
5. give students the idea that the lecture environment is safe by indicating that they are free to ask questions or make comments at any time and that these are welcomed because you prefer the communication be multidirectional
6. discuss the use of office hours, tutorials, discussion (or quiz) sections, and the laboratory sections
7. ensure that students have a good understanding of what is required to achieve a given grade in the course
8. if she/he is using a non-traditional lecture approach, it should be explained in detail how it works and why it was chosen
9. introduce the text, its features, and how it should be used
10. any remaining time can be used to begin presenting the course material, to perform some engaging demonstrations, or to show a short film or video

After each of the above items is presented, there should be a short pause for questions or comments from the students.

The Remaining Lectures

The remaining lectures are whatever you have planned for them to be. Since the various topics in physics are not independent of each other, but rather built upon common principles, lectures should begin with a reminder of what was covered in the previous lecture. At the end of the lecture, I even tell the students what will occur in the next lecture. This provides a seamless presentation/exploration of physics.

It's up to the instructor to keep the course moving at a reasonable pace. The pace should not be so fast that students get frustrated trying to write everything into their notes. Students tend to write everything that is written on a blackboard or on an overhead transparency into their notes. No thinking is required, just copying. While I'm on the topic student notes, most students do not take notes when points are made during demonstrations or simulations. Also, if students are asked to take notes during a film/video, they almost never write down anything that the instructor would consider important such as the process that led to a discovery or the evidence that supports a conclusion, or even the conclusion itself. Instead, they focus on trivia such as names, places, and dates.

We all know that attention spans are short. Patience tends to be short as well. The best way to keep minds from wondering is to periodically switch between any of the following modes: copying notes, giving examples, doing a demonstration, looking at a computer simulation, getting the students to work out a problem, asking and answering questions, telling a humorous anecdote, etc. One can use Chapter 5 to time, plan, and organize lectures.

There's so much more to teaching than could ever be written here or anywhere else. Effective teaching, like "doing physics," is a skill that's acquired through study, practice, and performance. How can we expect any more from our students than we are willing to do ourselves?

CHAPTER 2

Physics Resources on the Internet

Most people have only been aware of the existence of the vast worldwide connection of computer networks called the internet for the last seven to nine years. The military and scientific communities developed the network for communication and data exchange beginning on September 2, 1969 when two computers at the University of California – Los Angeles communicated via a 4.5-m cable. The original idea was to develop a decentralized computer network that could continue to operate in the event of nuclear attack. Many consider the true birth of the internet to have been on January 2, 1983 when 400 computers first utilized the standard data transfer protocol called TCP/IP. Today, anyone with a computer, a modem, and only small amount of training can access the internet and find information on virtually any topic. Physics-related information includes simulation software, electronic journal articles, specialized e-mail discussion groups, whole course offerings, research news, and much, much more. After a brief introduction to the various aspects of the internet, information will be given on how to access it all.

Introduction to The World Wide Web (www)

The world wide web is a subset of the internet and may be defined as the global collection of linked information (hypertext) that is accessed via a common user interface. The most popular user interfaces are the browsers that include Netscape and Microsoft's Internet Explorer. The browser application is run on the user's computer after a modem or ethernet connection is initiated between the computer and the network via an internet provider. The internet provider could be your college or university, a local commercial service, or a nationwide commercial service (such as America Online). Depending on the service support available from your internet provider, gathering all of the necessary software (usually at no cost), configuring all of the software settings, and subsequently accessing the internet for the first time can be either simple or extremely frustrating. However, once everything has been set-up properly, connecting can be as simple as a single (or double) mouse click.

After connecting and starting the browser, one can then access information by typing in a URL (uniform resource locator) or by using a search engine to find information about a given topic. The URL address, illustrated in Figure 2-1, is made up of several parts and must be typed exactly to reach the desired information. This example URL is for the University of Wisconsin Physics Demonstration Page. The protocol for www documents is usually "http." Other protocols include: "ftp," "news," and "telnet." The domain gives the name of the computer (webphysics.ph), the university's name (msstate), and the type of institution (edu). The directory path is the specific part of the web server where the file is located. The document filename can be the name of a file that may be an HTML document, a picture, a sound file, or a video file. HTML (hypertext markup language) is the language of the internet. It is written as a text file and includes information concerning the format of the web page, text to appear on the page, links to audio/visual resource files included on the page, and URLs to other pages on the net.

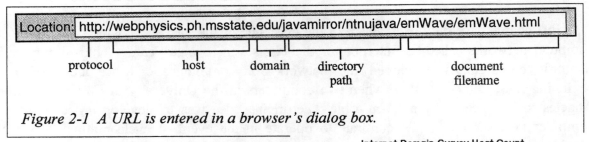

Figure 2-1 A URL is entered in a browser's dialog box.

The internet has grown exponentially over the past decade. The Internet Software Consortium (http://www.isc.org/) internet domain survey of August 2002 estimates that there are now more than 160 million host computers. Their results are shown in the graph to the right. The survey method changed in the mid-1990s and that is reflected on the graph.

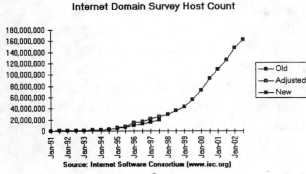

Notes about... URLs

- Once a page has been accessed, the URL can be saved by the browser as a bookmark (or favorite, etc.), so that one can revisit the page at some future time.
- The www is a rapidly changing medium; and therefore, URLs can become outdated. Usually, the person maintaining the web pages will provide a link to a new location. If a URL is given with a directory path and filename, but the document has been removed and has become unavailable, delete the filename and try again. If this fails, try deleting the directory path and the filename and access the main page of the domain. Sometimes, the main page will have a link to the information.
- Sometimes the directory path is shortened by the use of the tilde symbol, ~. On most keyboards, this symbol can be found in the upper, left-hand portion of the keyboard on the key above the tab key.

Search Engines and Portals

Virtually anything can be found on the internet. The trick is in how to find it. The exponential growth of the internet has also resulted in an exponential growth in advertising. Competition is tremendous to increase the number of visitors to a particular site. The first site you visit to search the web, to get stock quotes, to get the latest news, etc. is called a *portal*. Web browsers are pre-configured to automatically start at their respective portals. The browser can be reconfigured for the portal of your choice by entering the preferences window within the browser and making the change.

There are several search engines available that allow one to type a few, carefully chosen words to begin the search. Some search engine providers list categories for more casual browsing in a subject area. A few of the popular web search engines are listed in Table 2-1. Each search engine provider offers tips on maximizing efficiency in searching and gives information on using their particular search interface. Usually, the more specific one is, the easier it will be to find particular information. For example, if one enters "physics" into the search window, one can expect to find over ten million web pages that mention physics (and some that don't appear to have anything to do with physics as well). But if one enters "teaching physics" or "physics demonstrations," one will find a more practical number of pages.

Table 2-1. Selected Web Search Engines and Portals

Search Engine Provider	Web Address
AltaVista	www.altavista.com
Ask Jeeves	www.ask.com
Google	www.google.com
HotBot	www.hotbot.com
Netcenter	netcenter.com
Yahoo	www.yahoo.com

Notes about... Web Search Engines and Portals

- Often, the URL will be found by the search engine provider's software when it scans websites. In some cases, web pages are listed in the database of the search engine only after the author has submitted the URL to the search engine provider. Therefore, if a complete search is needed, two or more search engines should be used.

- If the particular organization you are looking for doesn't have a web page, you can sometimes find the information from a secondary source. For example, Jennifer is looking for the winter concert schedule for the Anytown Symphony Orchestra. A web search finds no link for the orchestra itself. Jennifer tries the following: "Anytown concert." The search engine finds that a radio station in Anytown has a webpage that lists all upcoming concerts in the area, including those for the orchestra and where tickets may be purchased. Jennifer successfully found the information she wanted.

E-mail and E-mail Discussion Groups

The process of sending a message to someone via electronic mail (or e-mail) is very similar to that of sending a letter to someone via surface mail. An e-mail program is used to compose, send, and receive messages. The message is usually a text file made up of a header and the body of the note. The header indicates the sender's and recipient's e-mail addresses and a subject line. The header may also indicate that "carbon copies" of the message were sent to others. Other files (images, documents, spreadsheets, compressed files, etc.) can be also attached or included with the note. Modern e-mail technologies allow text formatting, images, color, and sound to be added to what would otherwise be ordinary text.

E-mail addresses are similar to other web addresses in that the address is composed of a unique name and the domain name. For example, e-mail can be sent to the President of the United States by using the following address: president@whitehouse.gov. The name to the left of the "at" symbol, @, is some unique designator for the individual's mailbox at the domain.

If you have an e-mail program and an e-mail address, you can join an e-mail discussion group. The discussion group functions via a program called LISTSERV (or LISTPROC) designed to copy and distribute electronic mail to everyone who has joined the mailing list. E-mail discussion groups provide a unique opportunity to tap into the collective experience of, perhaps, hundreds of people with similar interests and concerns. For example, James composes a message regarding policies on the use of calculators in his calculus-based general physics course. He sends it to the discussion list's central address; and the LISTSERV automatically copies the message and sends it out to everyone on the list. At some later time, James checks for incoming e-mail and finds his message copied back to him along with any responses from the group regarding the use of calculators. He will also see messages *posted* by others and may choose to respond to those. Each topic in the on-going discussion is called a *thread*. Depending on the topic, threads can be a short as one message (the original and no responses) or may be carried on for several days or weeks. Often, a thread will evolve into a new topic as new issues are raised.

As of this writing, there are more than 300 000 e-mail discussion lists worldwide; and fortunately, there are search engines to find particular groups. Two useful search engines are:

    Catalist   http://www.lsoft.com/catalist.html
    Yahoogroups   http://groups.yahoo.com

Physics Resources on the Internet

A recent search with the word "physics" found more than 1600 physics-related groups. Many are private, some are public. To join a group at Yahoogroups, you can search and sign up from their website. E-mail from your groups may either be sent directly to you or read online.

Some of the discussion groups relevant to physics teaching are listed in Table 2-2. To subscribe to a list, send mail to the listserv administrator (the address on the left in the table) without a subject or a signature. In the body of the message write:

subscribe {name of list here} {your first name} {your last name}

For example, if I want to subscribe to the phys-l list, I would send this message to listserv@lists.nau.edu: subscribe phys-l David Marx. You can choose to receive all of the mail from the list only once per day by sending the following message to the listserv administrator (again with no subject or signature):

set {name of list here} **mail digest**

After you have subscribed, you would then send new messages to the address indicated on the right side of the table.

Table 2-2. Selected Physics-related E-mail Discussion Groups

physhare	Sharing Resources for High School Physics
	listserv@lists.psu.edu physhare@lists.psu.edu
	http://www.keithtipton.com/physhare/
	http://listserv.psu.edu/archives/physhare.html
phys-l	Forum for Physics Teachers
	listserv@lists.nau.edu phys-l@lists.nau.edu
	http://physicsed.buffalostate.edu/phys-l/
	http://lists.nau.edu/archives/phys-l.htm
PhysLrnR	Physics Learning Research
	listserv@listserv.boisestate.edu physlrnr@listserv.boisestate.edu
	http://listserv.boisestate.edu//archives/physlrnr.html
tap-l	Technical Aspects of Physics Labs and Lectures
	listserv@lester.appstate.edu tap-l@listserv.appstate.edu
	http://members.odyssey1.net/kctipton/faqs/tapfaq.html

Notes about...

Listserv Commands and e-Mail

Here are some other commands that you can also send to the list administrator:

- Help

  help

- Unsubscribe

  unsubscribe {name of list here}
  or
  signoff {name of list here}

- Temporarily pause your subscription

  set {name of list here} **mail postpone**

- To retrieve archived messages sent to the list

  get {name of list here} {name of list, again}.{YYMM}

  where the YY is the two digit year number (for 1995, YY = 95) and MM is the month (for November, MM = 11}. Note the period between the second listname and the date. If you need to find out how long the list has been archived, you can send the following message:

  index {name of list here}

- If there is a problem with the message you sent to the list administrator, you will get a message indicating the problem and the correct syntax.

- An etiquette has developed for using e-mail that has been dubbed *netiquette*. When you're conversing with any unknown group, it's easy to offend someone. Most people go out of their way to be civil, but occasionally arguments arise and people are said to be *flamed*. For a concise guide to netiquette, see

  http://www.fau.edu/netiquette/net/elec.html
  http://gloria-brame.com/glory/jour3.htm

- Veterans on the web often converse in a language, *cyberspeech*, that can be confusing to the novice. People will often use acronyms in e-mail messages to save keystrokes, to be civil, and to express emotions. People also add emotional expression to their writing by using a combination of ASCII characters to represent a facial expression. These combinations are called *emoticons*. For example, ;-) indicates that the writer is being sarcastic or mischievous by winking. The very first use of an emoticon was :-) by IBM neural network scientist Scott Fahlman on September 19, 1982. For a concise guide to emoticons, see

  http://www.cknow.com/ckinfo/emoticons.htm

Physics-related Resources on the World Wide Web

In this section you'll find several web sites that relate to physics and physics education. The URLs are listed in five categories: Organizations, Newsletters, & Journals; Physics Education; Web-based Courses & Lecture Notes; Lecture Demonstrations; Audio/Visual & Computer Simulations; and Miscellaneous. Each site has links to other places of interest, of course, so the ones listed below should be viewed as a starting point. Please note that these addresses were correct in January of 2003, but they are subject to change. Usually, when an address changes, the new URL is given, but not always. If the new address is not given, you may search for it using a search engine.

The web page for the Cutnell & Johnson text is

http://www.wiley.com/college/cutnell

Student Resources include:
**Interactive Learningware Problems
Selected Solutions to Homework Problems
Integration of Concepts Essays
Practice Quizzes for Each Chapter
Practice MCAT Quizzes
Practice Exams
Simulation Exercises
Interactive Illustrations
Web Links -** URLs for websites with corresponding material for each chapter of the text

Instructor Resources include:
WebCT – online course management
eGrade – online homework with randomized problem data for each student
WebAssign – online homework assignments
Web Links - URLs for websites with corresponding material for each chapter of the text

Web Resources: Organizations, Newsletters, and Electronic Journals

American Association for the Advancement of Science (AAAS)
   http://www.aaas.org/

American Association of Physics Teachers (AAPT)
   http://www.aapt.org/

American Institute of Physics (AIP)
   http://www.aip.org/

American Institute of Physics: AIP Center for History of Physics Newsletter
   http://www.aip.org/history/

American Journal of Physics (AJP)
   http://ojps.aip.org/ajp/

Computers in Physics / Computing in Science and Engineering
   http://ojps.aip.org/cipo
   http://ojps.aip.org/cise

Education Policy Analysis Archives
   http://olam.ed.asu.edu/epaa

Electronic Journal of Science Education
   http://unr.edu/homepage/jcannon/ejse/ejse.html

Institute of Physics
   http://www.iop.org/

New Scientist
   http://www.newscientist.com/

PEN: The American Institute of Physics Bulletin of Physics Education News
   http://www.aip.org/enews/pen/

Physics News Update
   http://www.aip.org/physnews/update/

Physics Today
   http://www.physicstoday.org

Science News
> http://www.sciencenews.org/

Science Week
> http://scienceweek.com/

Scientific American
> http://www.sciam.com/

The Physics Teacher
> http://www.aip.org/tpt/

Web Resources

Physics Education

Cooperative Group Problem Solving
> http://groups.physics.umn.edu/physed/Research/CGPS/CGPSintro.htm

Laboratory for Research in Physics Education (LRPE) at the University of Maine
> http://lrpe.umephy.maine.edu/PhysicsEducation

Harvard University (Peer Instruction)
> http://mazur-www.harvard.edu/Education/EducationMenu.html

Modeling Workshop Project
> http://modeling.la.asu.edu/modeling.html

Physics Education Group at the University of Washington
> http://www.phys.washington.edu/groups/peg

Physics Education Research Group (PERG) at the University of Maryland
> http://www.physics.umd.edu/perg/

Physics Education Research Group (PERG) at the Ohio State University
> http://www.physics.ohio-state.edu/~physedu/home.htm

Socratic Dialog-Inducing (SDI) Labs
> http://www.physics.indiana.edu/~sdi/

Workshop Physics
> http://physics.dickinson.edu/~wp_web/WP_homepage.html

For more physics education research homepages – use this excellent resource.
http://www.physics.umd.edu/perg/homepages.htm

Web Resources: Web-based Courses & Lecture Notes

College Physics for Students of Biology and Chemsitry - a algebra/trig-based hypertextbook for first year undergraduate physics students
http://www.rwc.uc.edu/koehler/biophys/contents.html

Computational Science Education Project
http://csep1.phy.ornl.gov/csep.html

CyberProf
www.howhy.com/home/

Electronic University: A Guide to Distance Learning
http://www.petersons.com/distancelearning/

Introductory Physics Notes - IUN/FYDE distance education program of the University of Winnipeg
http://theory.uwinnipeg.ca/physics/

Physical Sciences Resource Center
http://www.psrc-online.org/

Physics 2000 – an interactive journey through modern physics
http://www.colorado.edu/physics/2000/index.pl

Physics for Beginners
http://physics.webplasma.com/physicstoc.html

PhysicsWeb
http://physicsweb.org/

Univeristy of Florida's Physics 3054 - second semester web-based course
http://www.phys.ufl.edu/~phy3054/Welcome.html

Web Physics
http://webphysics.davidson.edu/

World Lecture Hall
http://www.utexas.edu/world/lecture/

Web Resources: Lecture Demonstrations

Brown University Physics Lecture Demonstrations
http://www.physics.brown.edu/Studies/Demo/

Physics Instructional Resource Association (PIRA) serves the needs of Physics Instructional Support Professionals through sharing ideas about demonstrations, laboratory activities, learning centers, and instructional resources in general.
http://physics.csufresno.edu/pirapub/default.html
http://www.wfu.edu/physics/pira/PIRAHome3.html

University of Maryland Physics Lecture Demonstration Facility Reference Materials
http://www.physics.umd.edu/deptinfo/facilities/lecdem/refs/refs.htm

University of Texas Physics Demonstration Page
http://www.physics.umd.edu/lecdem/

University of Wisconsin Physics Demonstration Page
http://demo1.physics.wisc.edu/

Wonders of Physics - physics demonstration shows and videos
http://sprott.physics.wisc.edu/wop.htm

Web Resources: Audio / Visual & Computer Simulations

Catholic Memorial Physics Demonstration page - JAVA applets for Mechanics, Wave Motion, Electricity and Magnetism and other demonstrations
http://www.cath-mem.org/physics/Demoes.htm

Crocodile Clips - simulation software for high school science includes mechanics, kinematics, sound, electricity, and electronics
http://www.crocodile-clips.com

Explore Science - multimedia demos for physics and other sciences
http://www.explorescience.com/

Fun Physics JAVA Applets
http://www3.adnc.com/~topquark/fun/fun.html

Interactive Physics Software - information concerning commercially available software
    http://www.interactivephysics.com

Interactive Science Activities on the Web
    http://www.bridgewater.edu/departments/physics/ISAW

Internet Software Archive
    http://www.download.com
    http://www.tucows.com/software.html

LivePhoto Physics Project
    http://physics.bsc.muskingum.edu/livephoto/

On Screen Particle Physics - information about the program
    http://www.onscreen-sci.com

Open Source Physics Education – a synergy of curriculum development, computational physics, and physics education research
    http://www.opensourcephysics.org/default.html

PBS: Nova Online
    http://www.pbs.org/wgbh/nova/

Physics Academic Software - commercial software programs
    http://webassign.net/pasnew/

Physics Software Packages
    http://www.saintmarys.edu/~rtarara/software.html

Physlets - physics applets
    http://webphysics.davidson.edu/Applets/Applets.html

Software Teaching of Modular Physics (SToMP)
    http://www.ph.surrey.ac.uk/stomp

VernierScientific Software
    http://www.vernier.com

WebPhysics - teaching and learning physics using world wide web technologies
    http://WebPhysics.davidson.edu

World in Motion
    http://members.aol.com/raacc/wim.html

Physics Resources on the Internet

Web Resources: Miscellaneous

Astronomy Education Resources
   http://www.ifa.hawaii.edu/tops/resources.html

Bad Physics - misconceptions people have concerning physics topics
   http://www.eskimo.com/~billb/miscon

The Directory of Phys-L/PhyShare Members' Webpages – for students and instructors
   http://w3.chrlmi.cablespeed.com/~exit60/phyweb.html

Donald Simanek's Science Links - numerous science links, many unusual
   http://www.lhup.edu/~dsimanek

Einstein Online - a comprehensive Albert Einstein website
   http://westegg.com/einstein/

Frank Potter's Science Gems - comprehensive list of more than 11 000 physical science education-related sites for all levels
   http://www.sciencegems.com/

Hands-on Centers Worldwide
   http://www.cs.cmu.edu/~mwm/sci.html

How Things Work
   http://howthingswork.virginia.edu/

Interactive Periodic Table
   http://www.webelements.com/

The Internet Pilot to Physics (TIPTOP) - international links, news, and software
   http://www.tp.umu.se/TIPTOP/

J-Track: Real Time Satellite Tracking from NASA
   http://liftoff.msfc.nasa.gov/RealTime/JTrack

National Aeronautics and Space Administration (NASA) Home Page
   http://www.nasa.gov

Net Advance of Physics - resources for introductory and advanced physics
   http://web.mit.edu/~redingtn/www/netadv/welcome.html

NIST Reference on Constants, Units, and Uncertainty
   http://physics.nist.gov/cuu/

Physical Sciences Resource Center
> http://www.psrc-online.org

The Physics e-Source
> http://www.dctech.com/physics

Physics Pavilion
> http://www.pen.k12.va.us/Pav/Science/Physics/

PhysicsWeb
> http://physicsweb.org/

Roller Coaster Physics: A Guide to Amusement Parks for Teachers
> http://www.pen.k12.va.us/Pav/Science/Physics/book/home.html

Shortcuts to Physics, Engineering, and Mathematics Departments
> http://chat.wcc.cc.il.us/~flemmerh/PEMDepts.htm

Society of Physics Students (SPS) & Sigma Pi Sigma (physics honor society)
> http://www.aip.org/education/sps/index.html

Unit Conversions - automatic unit converters
> http://digitaldutch.com/unitconverter/

The Why Files - an exploration of the science behind the news
> http://whyfiles.news.wisc.edu/

CHAPTER 3

Print, Audio/Visual, and Computer Resources

As the title implies, this chapter provides pedagogic resources in print as well as sources for films, videos, and computer software. The bibliographic information was adapted and updated from the extensive bibliography in J. Richard Christman's *Instructor's Manual* (for *Fundamentals of Physics, 5e.* by Halliday, Resnick, and Walker and published by John Wiley & Sons, New York, 1997). The bibliographic information is divided into the following categories:

> General Pedagogy Laboratory
> Computers in Physics Mechanics
> Thermodynamics Waves
> Electricity & Magnetism Light & Optics
> Modern Physics

The journals cited are:

American Journal of Physics	*AJP*
Computers in Physics	*CP*
International Journal of Science Education	*IJSE*
Journal of Computers in Mathematics and Science Education	*JCMSE*
Journal of Research on Computing in Education	*JRCE*
Journal of Research in Science Teaching	*JRST*
Physics Education	*PE*
The Physics Teacher	*TPT*
Research in Science and Technological Education	*RSTE*
Scientific American	*SA*

Publications and other materials from the AAPT can be obtained by contacting:

> American Association of Physics Teachers
> One Physics Ellipse
> College Park, MD 20740-3845
> (301) 209 - 3300
> http://www.aapt.org

The audio/visual and computer resources follow the print resource sections.

Print Resources: General Pedagogy

1. Arnold B. Arons, *Teaching Introductory Physics*, John Wiley & Sons, New York (1996).
2. Gardo Blado, "The MCAT Physics Test," *TPT* **38**, 364 (September 2000).
3. J. Bolton and S. Ross, "Developing Students' Physics Problem-solving Skills," *PE,* **32**, 176 (1997).
4. Juan Miguel Campanario, "Using Counterintuitive Problems in Teaching Physics," *TPT* **36**, 439 (1998).
5. Keith Devlin, "Rather than Scientific Literacy, Colleges should Teach Scientific Awareness," *TPT* **66**, 559 (July 1998).
6. R. Di Stefano, "The IUPP Evaluation: What We Were Trying to Learn and How We Were Trying to Learn It," *AJP*, **64**, 49 (January 1996).
7. R. Di Stefano, "Preliminary IUPP Results: Student Reaction to In-class Demonstrations and to the Presentation of Coherent Themes," *AJP*, **64**, 58 (January 1996).
8. L. W. DuBeck, S. E. Moshier, J. E. Boss, *Fantastic Voyages: Learning Science through Science Fiction Films*, AIP Press, Woodbury, New York (1994).
9. Kenneth Eble, *The Craft of Teaching: A Guide to Mastering the Professor's Art*, Jossey-Bass Publishers (1994).
10. Allen Feldman, "Enhancing the Practice of Physics Teachers: Mechanisms for the Generation and Sharing of Knowledge and Understanding in Collaborative Action Research," *JRST*, **33**, 513 (1996).
11. Dorothy L. Gabel, ed. *Handbook on Research on Science Teaching and Learning,* McMillan Publishing, New York (1994).
12. Ronald Gautreau and Lisa Novemsky, "Concepts First - A Small Group Approach to Physics Learning," *AJP*, **65**, 418 (1997).
13. Diane J. Grayson and Lillian McDermott, "Use Of The Computer For Research On Student Thinking In Physics," *AJP* **84**, 557 (May 1996).
14. Diane J. Grayson, "A Holistic Approach to Preparing Disadvantaged Students to Succeed in Tertiary Science Studies. Part II: Outcomes of the Science Foundation Programme," *IJSE*, **19**, 107 (1997).
15. Richard R. Hake, "Socratic Pedagogy in the Introductory Physics Laboratory," *TPT* **30**, 546 (December 1992).
16. David Hammer, "More than Misconceptions: Multiple Perspectives on Student Knowledge and Reasoning and an Appropriate Role for Education Research," *AJP*, **64**, 1316 (1996).
17. Maher Z. Hashweh, "Science Teachers' Epistemological Beliefs in Teaching," *JRST*, **33**, 47 (1996).

18. Patricia Heller, Ronald Keith, and Scott Anderson, "Teaching Problem Solving With Cooperative Grouping. Part 1: Group Versus Individual Problem Solving," *AJP* **60**, 627 (July 1992). Patricia Heller, Ronald Keith, and Scott Anderson, "Teaching Problem Solving With Cooperative Grouping. Part 2: Designing Problems And Structuring Groups," *AJP* **60**, 637 (July 1992).
19. Molly Johnson, "Facilitating High Quality Student Practice in Introductory Physics," *AJP* **69**, S2 (Issue S1, July 2001)
20. C. W. J. M. Klaassen and P. L. Lijnse, "Interpreting Students' and Teachers' Discourse in Science Classes: an Underestimated Problem?" *JRST*, **33**, 115 (1996).
21. P. W. Laws, "Promoting Active Learning Based on Physics Education Research in Introductory Physics Courses," *AJP*, **65**, 13 (1997).
22. William J. Leonard, Robert J. Dufresne, and Jose P. Mestre, "Using Qualitative Problem Solving Strategies to Highlight the Role of Conceptual Knowledge in Solving Problems," *AJP*, **64**, 1495 (1996).
23. Cedric J. Linder and Greg Hillhouse, "Teaching by Conceptual Exploration," *TPT*, **34**, 332 (1996).
24. David P. Maloney, Thomas L. O'Kuma, Curtis J. Hieggelke, and Alan van Heuvelen, "Surveying Students' Conceptual Knowledge of Electricity and Magnetism," *AJP* **69**, S12 (Issue S1, July 2001).
25. Richard P. McCall, "Physics and Pharmacy: More than 'Ph'," *TPT* **36**, 408 (October 1998).
26. Lillian C. McDermott, Peter S. Shaffer, and Mark L. Rosenquist, *Physics by Inquiry: An Introduction to Physics and the Physical Sciences*, John Wiley & Sons, New York (1995).
27. Lillian C. McDermott, "Millikan Lecture 1990: What we teach and what is learned - Closing the gap," *AJP* **59**, 301 (April 1991).
28. Lillian C. McDermott and Peter S. Shaffer, "Research As A Guide For Curriculum Development: An Example From Introductory Electricity. Part I: Investigation Of Student Understanding," *AJP* **60**, 994 (November 1992). Lillian C. McDermott and Peter S. Shaffer, "Research As A Guide For Curriculum Development: An Example From Introductory Electricity. Part II: Design of Instructional Strategies," *AJP* **60**, 1003 (November 1992). An erratum appears in *AJP* **61**, 81 (January 1993).
29. Lillian C. McDermott, Peter S. Shaffer, and Mark D. Somers, "Research As A Guide For Teaching Introductory Mechanics: An Illustration In The Context Of The Atwood's Machine," *AJP* **62**, 46 (January 1994).
30. Eric Mazur, *Peer Instruction: A User's Manual*, Prentice-Hall, New Jersey, 1997.
31. David E. Meltzer and Kandiah Manivannan, "Promoting Interactivity in Physics Lecture Classes," *TPT* **34**, 72 (February 1996).
32. John Mottman, "Innovations in Physics Teaching - A Cautionary Tale," *TPT* **37**, 74 (February 1999).

33. Michael Prosser, Paul Walker, and Rosemary Millar, "Differences in Students' Perceptions of Learning Physics," *PE*, **31**, 43 (1996).
34. Edward F. Redish, "Implications Of Cognitive Studies For Teaching Physics," *AJP* **82**, 796 (September 1994).
35. Robert H. Romer, "Reading The Equations And Confronting The Phenomena-- The Delights And Dilemmas Of Physics Teaching," *AJP* **61**, 128 (February 1993).
36. Rachel E. Scherr, "An Implementation of *Physics by Inquiry* in a Large –enrollment Class," *TPT* **41**, 113 (February 2003).
37. Beth Thaker, Eunsook Kim, Kelvin Trefz, and Suzanne M. Lea, "Comparing Problem Solving Performance Of Physics Students In Inquiry-Based And Traditional Introductory Physics Courses," *AJP* **82**, 627 (July 1994).
38. Alan Van Heuvelen, "Learning To Think Like A Physicist: A Review Of Research-Based Instructional Strategies," *AJP* **59**, 891 (October 1991).
39. Alan Van Heuvelen, "Overview: Case Study Physics," AJP **59**, 898 (October 1991).
40. Emily H. van Zee and James Minstrell, "Reflective Discourse: Developing Shared Understandings in a Physical Classroom," *IJSE*, **19**, 209 (1997).
41. Mark Vonracek, "Enhancing Student Learning by Tapping into Physics They Already Know," *TPT* **41**, 109 (February 2003).
42. Mike Watts and Keith S. Taber, "An Explanatory Gestalt of Essence: Students' Conceptions of the 'Natural' in Physical Phenomena," *IJSE*, **18**, 939 (1996).
43. Jack Wilson, ed. *Conference on the Introductory Physics Course*, John Wiley & Sons, New York, 1997.
44. Augden Windelborn, "Telepresent Teaching," *TPT* **38**, 16 (January 2000).

Laboratory

1. Arnold B. Arons, "Guiding Insight and Inquiry in the Introductory Physics Laboratory," *TPT* **31**, 278 (May 1993).
2. T. J. Bensky asnd S. E. Frey, "Computer Sound Card Assisted Measurements of the Acoustic Doppler Effect for Accelerated and Unaccelerated Sound Sources," *AJP* **69**, 1231 (December 2001).
3. Christopher G. Deacon, "Error Analysis in the Introductory Physics Laboratory," *TPT* **30**, 368 (September 1992).
4. A. Dupré and P. Janssen, "An Accurate Determination of the Acceleration of Gravity *g* in the Undergraduate Laboratory," *AJP* **68**, 704 (August 2000).
5. Ronald Edge, *String and Sticky Tape Experiments,* AAPT.

6. Robert Ehrlich, "Ruler Physics: Thirty-four Demonstrations Using A Plastic Ruler," AJP **62**, 111 (February 1994).
7. *Amusement Park Physics,* Carole Escobar, ed., AAPT, 1994.
8. Paul J. Germann and Roberta J. Aram, "Students Performances on the Science Processes of Recording Data, Analyzing Data, Drawing Conclusions, and Providing Evidence," *JRST*, **33**, 773 (1996).
9. Haym Kruglak, "Canned Physics, *TPT* **30**, 392 (October 1992).
10. Jim Malone and Dave Holzwarth, "A Laboratory-Based Final Exam in Mechanics," *TPT* **33**, 99 (February 1995).
11. Jim Malone, "Improving the Accuracy of Photogates," *TPT* **64**, 235 (1996).
12. Karl C. Mamola, ed. *Apparatus for Teaching Physics*, AAPT.
13. Wayne E. McGovern, "The Range Of A Data Set: Its Relationship To The Standard Deviation For Various Distributions," *AJP* **60**, 943 (October 1992).
14. R. C. Nicklin and Robert Miller, "Using Magnetic Switches in Motion Experiments," *TPT* 33, 118 (February 1995).
15. Douglas D. Osheroff, "The Nature of Discovery in Physics," *AJP* **69**, 26 (January 2001).
16. Stuart M. Quick, "A Computer-Assisted Free-Fall Experiment For The Freshman Laboratory," *AJP* **57**, 814 (September 1989).
17. Albert J. Read, "Hands-on Exhibits in Physics Education," *AJP* **57**, 393 (1989).
18. Eduardo E. Rodriguez, "A Proposal for Experimental Homework," *TPT* **36,** 435 (1998).
19. Fritz Schoch and Waiter Winiger, "How to Measure g Easily with $\sim 10^{-4}$ Precision in the Beginners' Lab," *TPT* **29**, 98 (February 1991).
20. David R. Sokoloff, Priscilla Laws, and Ronald Thornton, *Realtime Physics*, John Wiley & Sons, New York (1998). This is a set of manuals to be used in a lab with MBE tools. It is a powerful way to enhance conceptual learning.
21. Ronald K. Thornton and David R. Sokoloff, "Learning Motion Concepts Using Real-time Microcomputer-Based Laboratory Tools," *AJP* **58**, 858 (1990).
22. John Wessner, "The Average Bears Noting," *TPT* **31**, 548 (December 1993).

Computers in Physics

1. David Ayersman, "Reviewing the Hypermedia-based Learning Research," *JRCE,* **28**, 500 (1996).
2. Robert J. Beichner, "Applications of Macintosh Microcomputers in Introductory Physics," *TPT* **27**, 348 (May 1989).
3. R. E. Benenson and W. Bauer, "Frame Grabbing Techniques In Undergraduate Physics Education," *AJP* **61**, 848 (September 1993).

4. Marvin L. DeJong, "Computers in Introductory Physics," *CP* **5**, 12 (January/February 1991).
5. Denis Donnelly, "Equation-Solving Software Packages: Uses In The Undergraduate Curriculum," *AJP* **58**, 585 (June 1990).
6. Paula V. Engelhardt, Scott F. Schultz, John E. Gastineau, Margaret H. Gjertsen, and John S. Risley, "Teaching the Use of Spreadsheets for Physics," *TPT* **31**, 546 (December 1993).
7. Lawrence T. Escalada and Dean A. Zollman, "An Investigation on the Effects of Using Interactive Digital Video in a Physics Classroom on Student Learning and Attitudes, "*JRST*, **34**, 467 (1997).
8. Bat-Sheva Eylon, Miky Ronen, and Uri Ganiel, "Computer Simulations as Tools for Teaching and Learning: Using a Simulation Environment in Optics," *JSET*, **5**, 93 (1996).
9. Robert G. Fuller, *Computers in Physics Education*, AAPT.
10. Harvey Gould, *An Itroduction to Computer Simulation Methods: Applications for Physical Systems, 2e.*, Addison-Wesley Publ., 1996.
11. M. Graney and V. A. DiNoto, "Digitized Video Images as a Tool in the Physics Lab," *TPT* **33**, 460 (October 1995).
12. Rick Guglielmino, "Using Spreadsheets in an Introductory Physics Lab," *TPT* **27**, 175 (March 1989).
13. William G. Harter, "Nothing Going Nowhere Fast: Computer Graphics in Physics Courses," *CP* **5**, 466 (Sep/Oct 1991).
14. R. B. Hicks and H. Laue, "A Computer-Assisted Approach To Learning Physics Concepts," *AJP* **57**, 807 (September 1989).
15. Lorella M. Jones and Dennis J. Kane, "Student Evaluation Of Computer-Based Instruction In A Large University Mechanics Course," *AJP* **62**, 832 (1994).
16. E. Kashy, B.M. Sherrill, Y. Tsai, D. Thaler, D. Weinshank, M. Engelman, and D. J. Morrissey, "CAPA--An Integrated Computer-Assisted Personalized Assignment System," *AJP* **61**, 1124 (December 1993).
17. E. Kashy, S.J. Gaff, N. H. Pawley, W.L. Stretch, and S. L. Wolfe, "Conceptual Questions In Computer-Assisted Assignments," *AJP* **63**, 1000 (1995).
18. Michael E. Krieger and James H. Stith, "Spreadsheets in the Physics Laboratory," *TPT* **28**, 378 (May 1990).
19. William M. MacDonald, Edward F. Redish, and Jack M. Wilson, "The M.U.P.P.E.T. Manifesto," *CP* **2**, 23 (July/August 1986).
20. Daryl W. Preston and R.H. Good, "Computers In The General Physics Laboratory," *AJP* **64**, 766 (June 1996).
21. Edward F. Redish and Jack M. Wilson, "Student Programming In The Introductory Physics Course: M.U.P.P.E.T.," *AJP* **61**, 222 (March 1993).
22. John S. Risley, "Using Physics Courseware," *TPT* **27**, 188 (March 1989).
23. Daniel J. Suson, Lionel D. Hewett, Jim McCoy, and Vaughn Nelson, "Creating a Virtual Physics Department." *AJP* **67**, 520 (1999)

24. Robert F. Tinker, "Computer-aided Student Investigations," *CP* **2**, 46 (January/February 1988).
25. David L. Wagner, "Using Digitized Video for Motion Analysis," *TPT* **32**, 240 (April 1994).
26. Herman Weller, "Assessing the Impact of Computer-based Learning in Science," *JRCE*, **28**, 461 (1996).

Print Resources

Mechanics

1. Cliff Frohlich, "Resource Letter PS-I: Physics of Sports," *AJP* **54**, 590 (July 1986).
2. *Physics of Sports*, Cliff Frohlich., ed., AAPT.
3. Ray G. Van Ausdal, "Structured Problem Solving in Kinematics," *TPT* **26**, 518 (November 1988).
4. *Connecting Time and Space*, Harry E. Bates, ed., AAPT, 1992.
5. Roger Cowley, "A Classroom Exercise To Determine The Earth-Moon Distance," *AJP* **57**, 351 (April 1989).
6. Heilbron, "The Politics Of The Meter Stick," *AJP* **57**, 988 (November 1989).
7. Wayne Itano and Norman F. Ramsey, "Accurate Measurement of Time," *SA* **269**, 56 (July 1993).
8. *The International System of Units*, 2nd ed., Robert A. Nelson, ed., AAPT, 1983.
9. Robert O'Keefe and Bahman Ghavimi-Alagha, "The World Trade Center And The Distance To The World's Center," *AJP* **60**, 183 (February 1992).
10. Mark A. Peterson, "Error Analysis By Simulation," *AJP* **59**, 355 (April 1991).
11. Stanislaw Bednarek, "Magnetic Track For Experiments In Mechanics," *AJP* **60**, 664 (July 1992).
12. Robert J. Beichner, "Testing Student Interpretation Of Kinematics Graphs," *AJP* **82**, 750 (August 1994).
13. M. G. Calkin, "The Motion Of An Accelerating Automobile," *AJP* **58**, 573 (June 1990).
14. John Childs, "A Quick Determination of g Using Photogates," *TPT* **32**, 100 (February 1994).
15. Bill Crummett, "Measurements of Acceleration Due to Gravity," *TPT* **28**, 291 (May 1990).
16. Fred M. Goldberg and John H. Anderson, "Student Difficulties with Graphical Representations of Negative Values of Velocity," *TPT* **27**, 254 (April 1989).
17. Laurence I. Gould and Harry Workman, "Air Track With A Distributed Infrared Detector System," *AJP* **58**, 739 (August 1988).

18. Edwin Falser, "Instantaneous Velocity: A Different Approach," *TPT* **29**, 394 (September 1991).
19. Randall D. Knight, "The Vector Knowledge of Beginning Physics Students," *TPT* **33**, 74 (February 1995).
20. Richard V. Mancuso, "The Physical Significance of 'Other' Roots in Basic Physics Problems," *TPT* **30**, 533 (December 1992).
21. Lillian C. McDermott, Mark L. Rosenquist, and Emily H. vanZee, "Student Difficulties In Connecting Graphs And Physics: Examples From Kinematics," *AJP* **55**, 503 (June 1987).
22. Hans Pfister and Priscilla Laws, "Kinesthesia-l: Apparatus to Experience 1-D Motion," *TPT* **33**, 214 (April 1995).
23. R. Ramirez-Bon, "An Interesting Problem Solved by Vectors," *TPT* **28**, 594 (December 1990).
24. Mark L. Rosenquist and Lillian C. McDermott, "A Conceptual Approach To Teaching Kinematics," *AJP* **55**, 407 (May 1987).
25. Wolfgang Rueckner and Paul Titcomb, "An Accurate Determination Of The Acceleration Of Gravity For Lecture Hall Demonstration," *AJP* **55**, 324 (April 1987).
26. H. David Sheets, "Communicating with Vectors," *TPT* **36**, 520 (December 1998).
27. Joseph O. West, "The Atwood Machine: Two Special Cases," *TPT* **37**, 83 (February 1999).

Projectile Motion

1. A. Anicin, "The Rattle In The Cradle," *AJP* **55**, 533 (June 1987).
2. Yoav Ben-Dov, "Why the Dart Always Hits," *TPT* **31**, 526 (December 1993).
3. Peter J. Brancazio, "The Physics of Kicking a Football," *TPT* **23**, 403 (October 1985).
4. Ronald A. Brown, "Maximizing the Range of a Projectile," *TPT* **30**, 344 (September 1992).
5. K. R. Brownstein, "A Simple Derivation Of Centripetal Acceleration," *AJP* **62**, 946 (October 1994).
6. Michael A. Day and Martin H. Walker, "Experimenting with the National Guard: Field Artillery Gunnery," *TPT* **31**, 136 (March 1993).
7. Carey S. Inouye and Eric W. T. Chong, "Maximum Range of a Projectile," *TPT* **30**, 168 (March 1992).
8. David Keeports, "Numerical Calculation of Model Rocket Trajectories," *TPT* **28**, 274 (May 1990).
9. William J. Leonard and William J. Gerace, "The Power of Simple Reasoning," *TPT* 34, 280 (May 1996).

10. William M. MacDonald, "The Physics of The Drive in Golf," *AJP* **59**, 213 (March 1991).
11. Bengt Magnusson and Bruce Tiemann, "The Physics of Juggling," *TPT* **27**, 584 (November 1989).
12. Emie McFarland, "How Olympic Records Depend On Location," *AJP* **54**, 513 (June 1986).
13. Jeffrey W. Schnick, "Projectile Motion Details," *TPT* **32**, 266 (May 1994).
14. A. Tan and A. C. Giere, "Maxima Problems In Projectile Motion," *AJP* **55**, 750 (August 1987).
15. Michael Volpe, "Super Bowl Physics," *TPT* **32**, 399 (October 1994).
16. James S. Walker, "Projectiles: Are They Coming or Going?" *TPT* **33**, 282 (May 1995).

Forces and Newton's Laws of Motion

1. James E. Court, "Free-Body Diagrams," *TPT* **31**, 104 (February 1993).
2. Edward A. Desloge, The Empirical Foundation Of Classical Dynamics, *AJP* **57**, 704 (August 1989).
3. Helen M. Doerr, "Experiment, Simulation, and Analysis: an Integrated Instructional Approach to the Concept of Force," *IJSE*, **19**, 265 (1997).
4. A. P. French, "On Weightlessness," *AJP* **63**, 105 (February 1995).
5. Colin Gauld, "Using Colliding Pendulums to Teach Newton's Third Law," *TPT* **37**, 116 (February 1999).
6. D. Green and David T. Hartney, "Newton's Truck: Determining Mass and Finding g by Pushing a Truck across a Parking Lot," *TPT* **26**, 448 (October 1988).
7. Georg Hähner and Nicholas Spenser, "Rubbing and Scrubbing," *Physics Today*, September 1998. This is a fairly complete look at friction on all scales.
8. R. R. Hake, "Interactive-engagement versus Traditional Methods: A Six-Thousand Student Survey of Mechanics Test Data for Introductory Physics Courses," *AJP* **66**, 64 (1998).
9. Brian Hasson and Amy L. R. Bug, "Hands-on and Computer Simulations," TPT **33**, 230 (April 1995).
10. David Hestenes, Malcolm Wells, and Gregg Swackhamer, "Force Concept Inventory," *TPT* **30**, 141 (March 1992). *Also see* Douglas Huffman and Patricia Heller, "What Does the Force Concept Inventory Actually Measure?," *TPT* **33**, 138 (March 1995), and the response by Hestenes and Kalloun in *TPT* **33**, 502 (November 1995), and the response to that article by Heller and Huffman in *TPT* **33**, 503 (November 1995).
11. Joseph B. Keller, "Newton's Second Law," *AJP* **55**, 1145 (December 1987).
12. Alexander S. Kondratyev and William Sperry, "Direct Use of Vectors in Mechanics Problems," *TPT* **32**, 416 (October 1994).

13. Douglas A. Kurtze, "Teaching Newton's Second Law - A Better Way," *TPT* **29**, 350 (September 1991).
14. Brian Lane, "Why Can't Physicists Draw FBDs?" *TPT* **31**, 216 (April 1993).
15. William Lone, "Novel Third-Law Demonstration," *TPT* **33**, 84 (February 1995).
16. Kathy Malone and Bob Reiland, "Exploring Newton's Third Law," *TPT* **33**, 410 (September 1995).
17. David P. Maloney, "Forces as Interactions," *TPT* **28**, 386 (September 1990).
18. Erwin Marquit, "A Plea For a Correct Translation of Newton's Law of Inertia," *AJP* **58**, 867 (September 1990).
19. John D. McGervey, "Hands-on Physics for Less than a Dollar per Hand," *TPT* **33**, 238 (April 1995).
20. Jonathan Mitschele and Matthew Muscato, "Demonstrating Normal Forces with an Electronic Balance," *TPT* **32**, 555 (December 1994).
21. Richard C. Morrison, "Weight and Gravity - the Need for Consistent Definitions," *TPT* **37**, 51 (1999).
22. Yee-kong Ng, Se-yuen Mak, and Choi-men Chung, "Demonstration of Newton's Third Law using a Balloon Helicopter," *TPT* **40**, 181 (March 2002).
23. Barbara Pecori and Giacomo Torzo, "Physics of the Seesaw," *TPT* **39**, 491 (November 2001).
24. Willard Sperry, "Placing the Forces on Free-Body Diagrams," *TPT* **32**, 353 (September 1994).
25. Michael Svonavec, "Accelerated Motion With A Variable Weight," *AJP* **55**, 753 (August 1987).
26. Alan Van Heuvelen, "Experiment Problems for Mechanics," *TPT* **33**, 176 (March 1995).
27. Yvette A. Van Hise, "Student Misconceptions in Mechanics: An International Problem," *TPT* **26**, 498 (November 1988).
28. R. E. Vermillion and G. O. Cook, "A Particle Sliding Down A Movable Incline: An Experiment," *AJP* **56**, 438 (1988).
29. Ronald A. Bryan, Robert Beck Clark, and Pat Sadberry, "Illustrating Newton's Second Law with the Automobile Coast-Down Test," *TPT* **26**, 442 (October 1988).
30. John H. Dodge, "There's More to It than Friction," *TPT* **29**, 56 (January 1991).
31. Richard V. Mancuso, "Quantitative Analysis of Moving Two Fingers Under a Meter stick," *TPT* **31**, 222 (April 1993).
32. Eugene E. Nalence, "Using Automobile Road Test Data," *TPT* **26**, 278 (May 1988).
33. Neil M. Shea, "Terminal Speed and Atmospheric Density," *TPT* **31**, 176 (March 1993).
34. Robert R. Speers, "Physics and Roller Coasters - The Blue Streak at Cedar Point," *AJP* **59**, 528 (June 1991).

35. Vassilis Stravinidis, "Demonstrating Normal Forces with Kitchen Scales," *TPT* **36**, 556 (December 1998).
36. Derek B. Swinson, "Physics and Skiing," *TPT* **30**, 458 (November 1992).
37. Pall Theodorsson, "A New Dynamics Cart on an Inclined Plane," *TPT* **33**, 458 (October 1995).
38. C. W. Tompson and J. L. Wragg, "Terminal Velocity on an Air Track," *TPT* **29**, 178 (March 1991).
39. William S. Wagner, "Automobile Deceleration Force By The Coast-Down Method," *AJP* **54**, 1049 (November 1986).
40. William M, Wehrbein, "Frictional Forces On An Inclined Plane," *AJP* **60**, 57 (January 1992).
41. Robert Weinstock, "The Heavier They Are, the Faster They Fall—An Elementary Rigorous Proof," *TPT* **31**, 56 (January 1993).
42. Karen Williams, "Inexpensive Demonstrator of Newton's First Law," *TPT* **38** (February 2000).
43. Metin Yersel, "A Simple Demonstration of Terminal Velocity," *TPT* **29**, 335 (September 1991).
44. Joseph M. Zayas, "Experimental Determination Of The Coefficient Of Drag Of A Tennis Ball," *AJP* **54**, 622 (July 1986).

Circular Motion

1. Paul D. Lee, "Circular Motion," *TPT* **33**, 49 (January 1995).
2. Channon P. Price, "Teacup Physics: Centripetal Acceleration," *TPT* **28**, 49 (January 1990).
3. Bill Wedemeyer, "Centripetal Acceleration-A Simpler Derivation," *TPT* **31**, 238 (April 1993).

Work and Energy

1. Arnold B. Arons, "Developing the Energy Concepts in Introductory Physics," *TPT* **27**, 506 (October 1989).
2. Dale R. Blaszczak, "The Roller Coaster Experiment," *AJP* **59**, 283 (March 1991).
3. Robert Ehrlich, "Using A Retractable Ball Point Pen To Test The Law Of Conservation Of Energy," *AJP* **64**, 176 (February 1996).
4. James K. Head, "Faith in Physics - Building New Confidence with a Classic Pendulum Demonstration," *TPT* **33**, 10 (January 1995).
5. Stan Jakuba, "Effect of Exercise Expressed in Joules and Watts," *TPT* **29**, 512 (November 1991).

6. Ronald A. Lawson and Lillian C. McDermott, "Student Understanding Of The Work-Energy And Impulse-Momentum Theorems," *AJP* **55**, 811 (September 1987).
7. A. John Mallinckrodt and Harvey S. Leff, "All About Work," *AJP* **60**, 356 (April 1992).
8. A. Sherwood and W. H. Bernard, "Work and Heat Transfer in the Presence of Sliding Friction," *AJP* **52**, 1001 (1984).
9. David G. Willey, "Conservation of Mechanical Energy Using a Pendulum," *TPT* **29**, 567 (December 1991).

Impulse-Momentum and Collisions

1. Nicholas E. Brown, "Impulsive Thoughts on Some Elastic Collisions," *TPT* **23**, 421 (October 1985).
2. Norman Derby, "Reality and Theory in a Collision," *TPT* **37**, 24 (1999).
3. Jason W. Dunn, "A Human Hamster Wheel?" *TPT* **36**, 545 (December 1998).
4. D. Easton, "Can a Fly Stop a Train?," *TPT* **25**, 374 (September 1987).
5. Ian R. Gatland, "Relative Speed in Elastic Collisions," *TPT* **33**, 98 (February 1995).
6. Margaret Stautberg Greenwood, "Conservation of Momentum and the Center of Mass of a Cart-Truck System," *TPT* **25**, 370 (September 1987).
7. Margaret Stautberg Greenwood, "Inclined Plane on a Frictionless Surface," *TPT* **28**, 109 (February 1990).
8. F. Hermann, "Demonstration Of A Slow Inelastic Collision," *AJP* **54**, 658 (July 1986).
9. W. Klein and G. Nimitz, "Inelastic Collision And The Motion Of The Center Of Mass," *AJP* **57**, 182 (February 1989).
10. A. McMath, "A Dynamics Cart Demonstration: Momentum, Kinetic Energy, and More," *TPT* **24**, 282 (May 1986).
11. Walter Roy Mellen, "Aligner for Elastic Collisions of Dropped Balls," *TPT* **33**, 56 (January 1995).
12. C. T. Tindle, "An Intuitive Approach to Collisions," *TPT* **36**, 344 (1998).
13. Bob Wade, "'Spiraling Back' After the Holidays," *TPT* **32**, 408 (1994).
14. Robert G. Watts and Steven Baroni, "Baseball Bat Collisions and the Resulting Trajectories of Spinning Balls," *AJP* **57**, 40 (January 1989).

Rotational Kinematics and Dynamics

1. Gordon J. Aubrecht, II, Anthony P. French, Mario Iona, and Daniel W. Welch, "The Radian-That Troublesome Unit," *TPT* **31**, 84 (February 1993).
2. Peter J. Brancazio, "Rigid-Body Dynamics of a Football," *AJP* **55**, 415 (1987).

3. Howard Brody, "The Moment of Inertia of a Tennis Racket," *TPT* **23**, 213 (April 1985).
4. H. Brody, "The Sweet Spot of a Baseball Bat," *AJP* **54**, 640 (July 1986).
5. Hans C. Ohanian, "Rotational Motion and the Law of the Lever," *AJP* **59**, 182 (February 1991).
6. Roger Blickensdefer, "The Wheel and the Galilean Transformation," *TPT* **26**, 160 (March 1988).
7. A. Domenech, T. Domenech, and J. Cebrian, "Introduction To The Study Of Rolling Friction," *AJP* **55**, 231 (March 1987).
8. J. E. Fredrickson, "The Tail-less Cat in Free-Fall," *TPT* **27**, 620 (1989).
9. John Ronald Galli, "Angular Momentum Conservation and the Cat Twist," *TPT* **33**, 404 (September 1995).
10. T. M. Kaalotas and A. R. Lee, "A Simple Device to Illustrate Angular Momentum Conservation and Instability," *AJP* **58**, 80 (January 1990).
11. Sol Krasner, "Why Wheels Work: A Second Version," *TPT* **30**, 212 (April 1992).
12. L. Lam and E. Lowry, "Static Friction of a Rolling Wheel," *TPT* **25**, 504 (November 1987).
13. James A. Lock, "An alternative approach to the teaching of rotational dynamics," *AJP* **57**, 428 (May 1989).
14. S. Y. Mak and K. Y. Wong, "A qualitative demonstration of the conservation of angular momentum in a system of two non-coaxial rotating disks," *AJP* **57**, 951 (October 1989).
15. Robert H. March, "Who will win the race?," *TPT* **26**, 297 (May 1988).
16. J. D. Nightingale, "Which Way Will the Bike Move?," *TPT* **31**. 244 (April 1993).
17. Raymond A. Serway, Jim Lehman, and Richard Hall, "The Ballistic Cart on an Incline Revisited," *TPT* **33**, 578 (December 1995).
18. Peter L. Tea, Jr., "Trouble on the loop-the-loop," *AJP* **55**, 826 (September 1987).
19. Peter L. Tea, Jr., "On seeing instantaneous centers of velocity," *AJP* **58**, 495 (May 1990).

Simple Harmonic Motion and Elasticity

1. James Casey, "The Elasticity of Wood," *TPT* **31**, 286 (May 1993).
2. Bernard J. Feldman, "What to Say about the Tacoma Narrows Bridge to Your Introductory Physics Class," *TPT* **41**, 92 (February 2003).
3. Thomas B. Greenslade, Jr. and Richard Wilcox, "An Inexpensive Young's Modulus Apparatus," *TPT* **31**, 116 (February 1993).
4. Richard B. Kidd and Stuart L. Fogg, "A Simple Formula for the Large-angle Pendulum Period," *TPT* **40**, 81 (February 2002).

5. Karl C. Mamola and Joseph T. Pollock, "The Breaking Broomstick Demonstration," *TPT* **31**, 230 (April 1993).
6. D. L. Mathieson, "The Tensile Strength of Paper," *TPT* **29**, 412 (September 1991).
7. Kenneth S. Mendelson, "Statics of ladder leaning against a rough wall," *AJP* **63**, 148 (February 1995).
8. Paul G. Menz, "The Physics of Bungee Jumping," *TPT* **31**, 483 (1993).
9. S. Porter, "Potential Energy of a Vertical Oscillator," *TPT* **31**, 175 (March 1993).
10. Lawrence Ruby, "Equivalent Mass of a Coil Spring," *TPT* **38**, 140 (March 2000).
11. Vincent Santarelli, Joyce Carolla, and Michael Ferner, "A New Look at the Simple Pendulum," *TPT* **31**, 236 (April 1993).
12. Cindy Schwarz, "The Not-So-Simple Pendulum," *TPT* **33**, 225 (April 1995).
13. Michael C. Wittman, Richard N. Steinberg, and Edward F. Redish, "Making Sense of How Students Make Sense of Mechanical Waves," *TPT* **37**, 15 (1999).

Fluids

1. Henry S. Bader and Costas E. Synolakis, "The Bernoulli - Poiseuille Equation," *TPT* **27**, 598 (November 1989).
2. Robert P. Bauman, "Archimedes' Bath," *TPT* **25**, 162 (March 1987).
3. Robert P. Bauman and Rolf Schwaneberg, "Interpretation of Bernoulli's Equation," *TPT* **32**, 478 (November 1994).
4. Daniel E. Beeker, "Depth Dependence of Pressure," *TPT* **28**, 486 (October 1990).
5. Mario Capitolo, "Phase-change Demonstration - Instant Gratification," *TPT* **36**, 349 (1998).
6. Harold Cohen and David Horvath, "Two Large-scale Devices for Demonstrating a Bernoulli Effect," *TPT* **41**, 9 (January 2003).
7. Ronald M. Cosby and Douglas E. Petry, "Simple Buoyancy Demonstrations Using Saltwater," *TPT* **27**, 550 (October 1989).
8. Thomas Bruce Daniel, "Archimedes' Principle without the King's Crown," *TPT* **36**, 557 (December 1998).
9. Samuel Derman, "A Pointed Demonstration of Surface Tension," *TPT* **29**, 414 (September 1991).
10. Charles N. Eastlake, "An Aerodynamicists View of Lift, Bernoulli, and Newton," *TPT* **40**, 166 (March 2002).
11. John N. Fox, Jerry K. Eddy, and Norman W. Gaggini, "A real-time demonstration of the depth dependence of pressure in a liquid," *AJP* **56**, 620 (July 1988).

12. Eric Kineanon, "Explanation of a Buoyancy Demonstration," *TPT* **33**, 31 (January 1995).
13. Dean O. Kuethe, "Confusion about Pressure," *TPT* **29**, 20 (January 1991).
14. Alan L. Lehman and Thomas A. Lehman, "An illustration of buoyancy in the horizontal plane," *AJP* **56**, 1046 (November 1988).
15. S. Y. Mak and K. Y. Wong, "The measurement of surface tension by the method of direct pull," *AJP* **58**, 791 (August 1990).
16. Walter Roy Mellen, "Oscillations of Eggs and Things," *TPT* **32**, 474 (November 1994).
17. Ellis Noll, "Confronting the Buoyant Force," *TPT* **40**, 8 (January 2002).
18. Costa Emmanuel Synolakis and Henry S. Badeer, "On combining the Bernoulli and Poiseuille equation -- A plea to authors of college physics texts," *AJP* **57**, 1013 (November 1989).
19. A. Tan, "The Shape of Streamlined Tap Water Flow," *TPT* **23**, 494 (November 1985).
20. R. E. Vermillion, "Derivations of Archimedes' Principle," *AJP* **59**, 761 (August 1991).
21. Chris Waltham, "Flight without Bernoulli," *TPT* **36**, 457 (1998).
22. David A. Ward, "Finding the Buoyant Force," *TPT* **32**, 114 (February 1994).
23. Klaus Weltner, "A comparison of explanations of the aerodynamic lifting force," *AJP* **55**, 5D (January 1987).
24. Klaus Weltner, "Aerodynamic Lifting Force," *TPT* **28**, 78 (February 1990).
25. Klaus Weltner, "Bernoulli's Law and Aerodynamic Lifting Force," *TPT* **28**, 84 (February 1990).

Thermodynamics

1. Russell Akridge, "Particle-Model Derivation of PV^γ," *TPT* **37**, 110 (February 1999).
2. Marcelo Alonso and Edward J. Finn, "An Integrated Approach to Thermodynamics in the Introductory Physics Course," *TPT* **33**, 296 (May 1995).
3. Richard A. Bartels, "Do darker objects really cool faster?," *AJP* **58**, 244 (March 1990).
4. Ralph Baierlein, "How entropy got its name," *AJP* **60**, 1151 (December 1992).
5. Philip O. Berge, G. Ulrich Nienhaus, and Jeffrey B. Ziegler, "Constant volume gas thermometer without mercury," *AJP* **62**, 666 (July 1994).
6. Charles H. Bennett, "Demons, Engines and the Second Law," *SA* **257**, 108 (November 1987).

7. Richard J. Bohan and Guy Vandegrift, "Temperature-driven Covection" *TPT* **41**, 76 (February 2003).
8. Ronald Bryan. "Avogadro's Number and the Kinetic Theory of Gases," *TPT* **38** (February 2000).
9. Manfred Bucher and Hugh A. Williamson, "Conversion of Temperature Scales," *TPT* **24**, 288 (May 1986).
10. Manfred Bucher, "Diagram of the second law of thermodynamics," *AJP* **81**, 462 (May 1993). William H. Cropper, "Carnot's Function: Origins of the Thermodynamic Concept of Temperature," *AJP* **55**, 120 (February 1987).
11. Peter Drago, "Teaching with Spreadsheets: An Example from Heat Transfer," *TPT* **31**, 316 (May 1993).
12. Hasan Fakhruddin, "Thermal Expansion 'Paradox'", *TPT* **31**, 214 (April 1993).
13. John N. Fox, "Measurement of thermal expansion coefficients using a strain gauge," *AJP* **58**, 875 (September 1990).
14. Hans U. Fuchs, "Entropy in the teaching of introductory thermodynamics," *AJP* **55**, 215 (March 1987).
15. Louie A. Galloway, III and John F. Wilson, Jr., "Measuring the Mechanical Equivalent of Heat-Electrically," TPT 30, 504 (November 1992).
16. Paul Hewitt, "Melting Ice," *TPT* **41**, 8 (January 2003).
17. Paul Inscho, "Mechanical Equivalent of Heat," *TPT* **30**, 372 (September 1992).
18. Dragia T. Ivanov, "Measuring the Speed of Molecules in a Gas," *TPT* **34**, 278 (May 1996).
19. William L. Kerr and Donald S. Reid, "Thermodynamics and Frozen Foods," *TPT* **31**, 52 (January 1993).
20. D. W. Lamb and G. M. White, "Apparatus to Measure Adiabatic and Isothermal Processes," *TPT* **34**, 290 (May 1996).
21. Bernard H. Lavenda, "Brownian Motion," *SA* **252**, 70 (February 1985).
22. Ed Lint, "Understanding Latent Heat of Vaporization," *TPT* **33**, 294 (1995).
23. Thomas V. Marcella, "Entropy Production and the Second Law of Thermodynamics: An Introduction to Second Law Analysis," *AJP* **60**, 888 (October 1992).
24. R. Mostert, "Classroom Experiments on Thermal Expansion of Solids," *TPT* **30**, 15 (January 1992).
25. George D. Nickas, "A thermometer based on Archimedes' principle," *AJP* **57**, 845 (September 1989).
26. Robert Otani and Peter Siegel, "Determining Absolute Zero in the Kitchen Sink," *TPT* **29**, 316 (May 1991).
27. W. G. Rees and C. Viney, "On cooling tea and coffee," *AJP* **56**, 434 (1988).
28. R. D. Russell, "Demonstrating Adiabatic Temperature Changes," *TPT* **25**, 450 (October 1987).
29. Levi Tansjp, "Comment on the Discovery of the Second Law," *AJP* **56**, 179 (February 1988).

30. Volker Thomson, "Response Time of a Thermometer," *TPT* **36**, 540 (December 1998).
31. Neff Weber, "Measuring the Mechanical Equivalent of Heat-Mechanically," *TPT* **30**, 507 (November 1992).
32. Klaus Weltner, "Measurement of specific heat capacity of air," *AJP* **61**, 661 (July 1993).
33. Shelley Yeo and Marjan Zadnik, "Introductory Thermal Concept Evaluation: Assessing Students' Understanding," *TPT* **39**, 496 (November 2001).

Print Resources: Wave Motion

1. Jan Paul Dabrowski, "Speed of Sound in a Parking Lot," *TPT* **28**, 410 (September 1990).
2. Charlotte Farrell, "A Sound Wave Demonstrator - For about $10," *TPT* **29**, 185 (March 1991).
3. Michael T. Frank and Edward Kluk, "Velocity of Sound in Solids," *TPT* **29**, 246 (April 1991).
4. Thomas B. Greenslade, Jr., "Experiments with Ultrasonic Transducers," *TPT* **32**, 392 (October 1994).
5. Donald E. Hall, "Sacrificing a Cheap Guitar in the Name of Science," *TPT* **27**, 673 (December 1989).
6. Brian Holmes, "The Helium-Filled Organ Pipe," *TPT* **27**, 218 (March 1989).
7. Robert G. Hunt, "Dancing to a Different Tune," *TPT* **31**, 206 (April 1993).
8. I. D. Johnston, "Standing waves in air columns: Will computers reshape physics courses?" *AJP* **61**, 996 (November 1993).
9. G. B. Karshner, "Direct method for measuring the speed of sound," *AJP* **57**, 920 (October 1989).
10. M. G. Raymer and Stan Micklavzina, "Demonstrating Sound Impulses in Pipes," *TPT* **33**, 183 (March 1995).
11. Thomas D. Rossing, "Resource Letter MA: Musical acoustics," *AJP* **55**, 589 (July 1987).
12. Thomas D. Rossing, ed., *Musical Acoustics,* AAPT.
13. Thomas D. Rossing, Daniel A. Russell, and David E. Brown, "On the acoustics of tuning forks," *AJP* **60**, 620 (July 1992).
14. Wolfgang Rueckner, Douglass Goodale, Daniel Rosenberg, Simon Steel, and David Tavilla, "Lecture Demonstration of Wine Glass Resonances," *AJP* **61**, 184 (February 1993).
15. Vincent Santarelli, Joyce Carolla, and Michael Ferner, "Standing Waves in a Mailing Tube," *TPT* **31**, 557 (December 1993).

16. Donald E. Shult, "Tone Holes and the Frequency of Open Pipes," *TPT* **29**, 16 (January 1991).
17. Nilgun Sungar, "Teaching the Superposition of Waves," *TPT* **34**, 236 (April 1996).
18. C. T. Tindle, "Pressure and Displacement in Sound Waves," *AJP* **54**, 749 (August 1986).
19. Frank Munley, "Phase and Displacement in Sound Waves," *AJP* **58**, 1144 (December 1988).
20. C. T. Tindle, "Decibels Made Easy," *TPT* **34**, 304 (May 1996).
21. Loren M. Winters, "A Visual Measurement of the Speed of Sound," *TPT* **31**, 284 (May 1993). Herbert T. Wood, "Mechanical Analogue of the Doppler Effect," *TPT* **30**, 340 (September 1992).
22. Rand S. Worland and D. David Wilson, "The Speed of Sound in Air as a Function of Temperature," *TPT* **37**, 53 (January 1999).
23. Junry Wu, "Are sound waves isothermal or adiabatic?" *AJP* **58**, 694 (July 1990). Also see the comment by Pieter B. Visscher, *AJP* **59**, 948 (October 1991).
24. An Zhong, "An acoustic Doppler shift experiment with the signal-receiving relay," *AJP* **57**, 49 (January 1989).

Electricity & Magnetism

1. Gary Benoit and Mauri Gould, "A New Device for Studying Electric Fields," *TPT* **29**, 182 (March 1991).
2. Edgar A. Bering III, Arthur A. Few, and James R. Benbrook, "The Global Electric Circuit," *Physics Today* October 1998, 24.
3. Ena S. Bichsel, Brenda Wilson, and Wilhelmus J. Geerts, "Magnetic Domains of Floppy Disks and Phone Cards using Toner Fluid," *TPT*, **40**, 150 (March 2002).
4. Sarah Fay and Angela Portenga, "Hey You! Shut the Refrigerator Door!" *TPT* **36**, 336 (1998). This article provides notes on measuring W/kW•h values for everyday appliances as a classroom exercise.
5. Salvatore Ganci, "An electroscope discriminating the sign of charges," *AJP* **62**, 474 (May 1994).
6. Gordon R. Gore and William R. Gregg, "Three Inexpensive High-Voltage Electricity Demonstrations," *TPT* **30**, 400 (October 1992).
7. Randal Harrington, "Getting a Charge out of Transparent Tape." *TPT* **38** (January 2000).

8. Dragia T. Ivanov, "Another Way to Demonstrate Lenz's Law," *TPT* **38** (January 2000).
9. Peter Peering, "On Coulomb's Inverse Square Law," *AJP* **60**, 988 (November 1992).
10. John G. King, Philip Morrison, Phylis Morrison, and Jerome Pine, "ZAP! Freshman Electricity and Magnetism using Desktop Experiments: A Progress Report," *AJP* **60**, 973 (November 1992). Waiter Roy Melleni, "Inexpensive Fun with Electrostatics," *TPT* **27**, 86 (February 1989).
11. Martha Lietz, "A Potential Guass' Law Lab," *TPT* **38** (April 2000).
12. Wolfgang Rueckner, Douglass Goodale, Daniel Rosenberg, and David Tavilla, "Demonstration of Charge Conservation," *AJP* **63**, 90 (January 1995).
13. Andrew A. Ruether, "Smart Bubbles," *TPT* **33**, 279 (May 1995).
14. Josip Slisk and Arkady Krokhin, "Physics or Fantasy?," *TPT* **33**, 210 (April 1995).
15. Ludwik Kowalski, "A Short History of the SI Units in Electricity," *TPT* **24**, 97 (February 1986).
16. L. Kristjansson, "On the drawing of lines of force and equipotentials," *TPT* **23**, 203 (April 1985).
17. S. Rainson, G. Transtmer, and L. Viennot, "Students' Understanding of Superposition of Electric Fields," *AJP* **82**, 1026 (November 1994).
18. Ross L. Spencer, "Electric field lines near an oddly shaped conductor in a uniform electric field," *AJP* **56**, 510 (June 1988).
19. Gay B. Stewart, "Measuring the Earth's Magnetic Field Simply," *TPT* **38** (February 2000).
20. S. Törnkvist, K.-A. Pettersson, and G. Transtrijmer, "Confusion by Representation: On Student's Comprehension of the Electric Field Concept," *AJP* **61**, 335 (April 1993).
21. Alan Wolf, Stephen J. Van Hook, and Eric R. Weeks, "Electric field line diagrams don't work," *AJP* **64**, 714 (June 1996).
22. Zvi Geller and Esther Bagno, "Does Electrostatic Shielding Work Both Ways?," *TPT* **32**, 20 (January 1994).
23. Richard E. Berg, "Van de Graaff Generators: Theory, Maintenance, and Belt Fabrication," *TPT* **28**, 281 (May 1990).
24. Wayne A. Powers, "Still More on the Coulomb Potential," *AJP* **57**, 375 (April 1989).
25. Bedamati Das, Avik Ghosh, and Pinaki Gupta-Bhaya, "Electrostatic energy of a system of charges and dielectrics," *AJP* **63**, 452 (May 1995).
26. Marvin L. DeJong, "Graphing Electric Potential," *TPT* **31**, 270 (May 1993).
27. H. W. Fulbright, "Simple and inexpensive teaching apparatus for absolute measurement of voltage," *AJP* **61**, 896 (October 1993).
28. Robert D. Smith, "Electrostatic Field Plotting Apparatus," *AJP* **58**, 410 (April 1990).

29. A. P. French, "Are the Textbook Writers Wrong about Capacitors?," *TPT* **31**, 156 (March 1993).
30. B. L. Illman and G. T. Carlson, "Equal Plate Charges on Series Capacitors?," *TPT* **32**, 77 (February 1994).
31. Edwin A. Karlow, "Let's Measure the Dielectric Constant of a Piece of Paper!," *TPT* **29**, 23 (January 1991).
32. L. Kowalski, "A Myth about Capacitors in Series," *TPT* **26**, 286 (May 1988).
33. Abdeljalil Metiouli, Claude Brassard, Jude Levasseur, and Michel Lavoie, "The Persistence of Students' Unfounded Beliefs about Electrical Circuits: the Case of Ohm's Law," *IJSE*, **18**, 193 (1996).
34. Helene F. Perry, Randall S. Jones, and Gregory N. Derry, "Capacitors in Parallel and Series," *TPT* **29**, 348 (September 1991).
35. Charles A. Sawicki, "Inexpensive Demonstration of the Magnetic Properties of Matter," *TPT* **36**, 553 (December 1998).
36. Susan M. Stocklmayer and David F. Treagust, "Images of Electricity: How do Novices and Experts Model Electric Current?" *IJSE*, **18**, 163 (1996).
37. Paul J. H. Tjossem and Victor Cornejo, "Measurements and Mechanisms of Thomson's Jumping Ring," *AJP* **68**, 238 (March 2000).
38. Donald M. Trotter, Jr., "Capacitors," *SA* **259**, 86 (July 1988).
39. Constantino A. Utreras-Diaz, "Dielectric Slab in a Parallel-plate Condenser," *AJP* **56**, 700 (August 1988).
40. David E. Wilson, "A direct laboratory approach to the study of capacitors," *AJP* **57**, 630 (July 1989).
41. N. Yan and H. K. Wong, "Force between the plates of a parallel-plate capacitor," *AJP* **61**, 1153 (December 1993).
42. Robert P. Bauman and Saleh Adams, "Misunderstandings of Electric Current," *TPT* **28**, 334 (May 1990).
43. John G. King and A. P. French, "Using a Multimeter to Study an RC Circuit," *TPT* **33**, 188 (March 1995).
44. Colin Terry, "Black-Box Electrical Circuits," *TPT* **33** 386 (September 1995).
45. Rodney C. Cross, "Magnetic lines of force and rubber bands," *AJP* **57**, 722 (August 1989).
46. W. Herreman and R. Huysentruyt, "Measuring the Magnetic Force on a Current-Carrying Conductor," *TPT* **33**, 288 (May 1995).
47. S. Y. Mak and K. Young, "Floating metal ring in an alternating magnetic field," *AJP* **54**, 808 (September 1986).
48. Greg Boebinger, Al Passner, and Joze Bevk, "Building World-Record Magnets," *SA* **272**, 58 (June 1995).
49. Peter Heller, "Analog demonstrations of Ampere's law and magnetic flux," *AJP* **60**, 17 (January 1992).
50. W. Klein and Thomas Unkelbach, "The magnetic field of a current conducting wire," *AJP* **81**, 659 (July 1993).

51. C. E. Zaspel, "An inexpensive Ampere's law experiment," *AJP* **56**, 859 (September 1988).
52. A. W. DeSilva, "Magnetically imploded soft Drink can," *AJP* **62**, 41 (January 1994).
53. William Lone, "A novel demonstration of induced EMF," *AJP* **61**, 90 (January 1993).
54. R. C. Nicklin, "Faraday's law - Quantitative experiments," *AJP* **54**, 422 (May 1986).
55. Thomas D. Rossing and John R. Hull, "Magnetic Levitation," *TPT* **29**, 552 (December 1991).
56. W. M. Saslow, "Electromechanical implications of Faraday's law: A problem collection," *AJP* **55**, 986 (November 1987).
57. P. Seligmann, "The Earth's magnetic field - A new technique," *AJP* **55**, 379 (April 1987).
58. Pearce Williams, "Why Ampere did not discover electromagnetic induction," *AJP* **54**, 306 (April 1986).
59. Alfred S. Goldhaber and W. Peter Trower, eds., *Magnetic Monopoles*, AAPT, 1991.
60. Alfred S. Goldhaber, "Resource Letter MM-1: Magnetic Monopoles," *AJP* **58**, 429 (May 1990).
61. Kenneth A. Hoffman, "Ancient Magnetic Reversals: Clues to the Geodynamo," *SA* **258**, 76 (May 1988).
62. Chin-Shan Lue, "A Direct Method of Viewing Ferromagnetic Phase Transition," *TPT* **32**, 304 (May 1994).
63. S. K. Runcorn, "The Moon's Ancient Magnetism," *SA* **257**, 60 (December 1987).
64. Deborah Schurr and Tim Usher, "Demonstrating Hysteresis in Ferroelectric Materials," *TPT* **33**, 30 (January 1995).

Print Resources

Light & Optics

1. Lorenzo Basano and Pasquale Ottonello, "Interference Fringes from Stabilized Diode Lasers," *AJP*, **68**, 345 (March 2000).
2. Harry E. Bates, Resource Letter RMSL-1: Recent measurements of the speed of light and the redefinition of the meter, AJP 58, 682 (August 1988).
3. Craig F. Bohren, "Multiple scattering of light and some of its observable consequences," *AJP* **55**, 524 (June 1987).
4. Paul Chagnon, "Animated Displays IV: Linear Polarization," *TPT* **31**, 489 (November 1993).

5. Adolf Cortel, "Demonstrating the Relationship between Energy and Frequency of Light," *TPT* **38**, (March 2000).
6. Melissa Dancy, Wolfgang Christian, and Mario Belloni, "Teaching with Physlets: Examples from Optics," *TPT* **40**, 494 (November 2002).
7. R. Davies, "Polarized Light Corridor Demonstrations," *TPT* **28**, 464 (October 1990).
8. Tom Donohue and Howard Wallace, "Ultraviolet Viewer," *TPT* **31**, 41 (1993).
9. Scott C. Dudley, "How to Quickly Estimate the Focal Length of a Diverging Lens," *TPT* **37**, 94 (February 1999).
10. Thomas B. Greenslade, "Quick Experiment on Reflection from Concave Mirrors," *TPT* **38** (April 2000).
11. James E. Kettler, "Index of Refraction by Reflected and Refracted Rays," *TPT* **32**, 190 (March 1994).
12. K. Koo, C. S. Chong, and Pen K. Merican, "Polarized light, Polaroid and your digital watch," *AJP* **83**, 184 (February 1995).
13. Robert W. Lawrence, " Magnification Ratio and the Lens Equations," *TPT* **38** (March 2000).
14. A. F. Leung, "Wavelength of light in water," *AJP* **54**, 956 (October 1986).
15. Alfred F. Leung, Simon George, and Robert Doebler, "Refractive Index of a Liquid Measured with a He-Ne Laser," *TPT* **29**, 226 (April 1991).
16. Alfred F. Leung and Simon George, "Simple Homemade Container for Measuring Refractive Index," *TPT* **30**, 438 (October 1992).
17. F. Melton, "A Surprising Demonstration of Total Internal Reflection," *TPT* **29**, 539 (November 1991).
18. Antonio B. Nassar, "Apparent Depth," *TPT* **32**, 526 (December 1994). For some comments on this paper, see *TPT* **33**, 198 (April 1995).
19. P. J. Ouseph, "Apparatus for Teaching Physics: Polarization of Reflected Light," *TPT* **40**, 438 (October 2002).
20. P. J. Ouseph, Kevin Driver, and John Conklin, "Polarization of Light by Reflection and the Brewster Angle," *AJP* **69**, 1166 (November 2001).
21. Leonard Parsons, "As easy as R, G, B," *TPT* **36**, 347 (1998).
22. Joseph L. Spradley, "Hertz and the Discovery of Radio Waves and the Photoelectric Effect," *TPT* **26**, 492 (November 1988).
23. Richard Atneosen and Richard Feinberg, "Learning optics with optical design software," *AJP* **59**, 242.
24. Ermanno F. Borra, "Liquid Mirrors," *SA* **270**, 76 (February 1994).
25. John W. W. Burrows, "Derivation of the mirror equation," *AJP* **54**, 432 (May 1986).
26. Samuel Derman, "Ray Tracing with Spherical and Parabolic Reflectors," *TPT* **28**, 590 (December 1990).
27. Pietro Ferraro, "Optics with Balloons," *TPT* **34**, 274 (May 1996).
28. Igal Galili, Fred Goldberg, and Sharon Bendall, "Some Reflections on Plane Mirrors and Images," *TPT* **29**, 471 (October 1991).

29. Igal Galili and Fred Goldberg, "Left-Right Conversions in a Plane Mirror," *TPT* **31**, 463 (November 1993).
30. Igal Galili, "Students' Conceptual Change in Geometrical Optics," *IJSE*, **18**, 847 (1996).
31. Igal Galili and Fred Goldberg, "Using a linear approximation for single-surface refraction to explain some virtual image phenomena," *AJP* **64**, 256 (March 1996).
32. Fred M. Goldberg and Lillian C. McDermott, "Student Difficulties in Understanding Image Formation by a Plane Mirror," *TPT* **24**, 472 (November 1986).
33. Fred M. Goldberg and Lillian C. McDermott, "An investigation of student understanding of the real image formed by a converging lens or concave mirror," *AJP* **55**, 108 (February 1987).
34. Diane J. Grayson, "Many Rays Are Better than Two," *TPT* **33**, 42 (January 1995).
35. John W. Hardy, "Adaptive Optics," *SA* **270**, 60 (June 1994).
36. Malcolm R. Howells, Janos Kirz, and David Sayre, "X-ray Microscopes," *SA* **264**, 80 (February 1991).
37. Jay S. Huebner and Terry L. Smith, "Why Magnification Works," *TPT* **32**, 102 (February 1994).
38. David R. Lapp, "Determining Plane Mirror Image Distance from Eye Charts," *TPT* **31**, 59 (January 1993).
39. Stuart Leinoff, "Ray Tracing with Virtual Objects," *TPT* **29**, 275 (May 1991).
40. Gordon P. Ramsey, "Reflective Properties of a Parabolic Mirror," *TPT* **29**, 240 (April 1991).
41. Ralph Baierlein and Vacek Miglus, "Illustrating double-slit interference: Yet another way," *AJP* **59**, 857 (September 1991).
42. Kai-yin Cheung and Se-yuen Mak, "Giant Newton's Rings," *TPT* **34**, 35 (January 1996).
43. William Moebs and Jeff Sanny, "A Simple Description of Coherence," *TPT* **32**, 54 (January 1994).
44. Glen M. Robinson, David M. Ferry, and Richard W. Peterson, "Optical Interferometry of Surfaces," *SA* **265**, 66 (July 1991).
45. Scott Whiteray, "Interference Patterns and Landing Aircraft," *TPT* **33**, 82 (February 1995).
46. Simon George and Robert Doebler, "Index of Refraction Using a Diffraction Pattern in a Fish tank," *TPT* **29**, 462 (October 1991).
47. James E. Kettler, "The compact disk as a diffraction grating," *AJP* **59**, 367 (April 1991).
48. F. Leung and Simon George, "Diffraction intensity of a phase grating submerged in different liquids," *AJP* **57**, 854 (September 1989).
49. Kenneth S. Mendelson and Frank G. Karioris, "Multislit diffraction patterns using a slit-grating combination," *AJP* **63**, 53 (January 1995).

50. Christian Nijldeke, "Compact Disc Diffraction," *TPT* **28**, 484 (October 1990).
51. Saeid Rahimi and Robert A. Baker, "Three-dimensional Display of Light Interference Patterns," *AJP* **67**, 453 (May 1999).
52. Philip Sadler, "Projecting Spectra for Classroom Demonstrations," *TPT* **29**, 423 (October 1991).
53. Mu-Shiang Wu and Shu-Ming Yang, "Dispersion and resolving power of a grating," *AJP* **54**, 735 (August 1986).
54. Vittorio Zanetti and John Harris, "Spectra of Three Light Sources with a CD," *TPT* **31**, 82 (February 1993).

Modern Physics

1. Ralph Baierlein, "Teaching $E = mc^2$," *AJP* **57**, 391 (May 1989).
2. Robert P. Bauman, "Mass and Energy: The Low-Energy Limit," *TPT* **32**, 340 (September 1994).
3. P. Bough, "The case of the identically accelerated twin," *AJP* **57**, 791 (September 1989).
4. Lisa N. Daly and George K. Horton, "The Universality of Time Dilation and Space Contraction," *TPT* **32**, 306 (May 1994).
5. L. Fadner, "Did Einstein really discover $E = mc^2$?," *AJP* **56**, 114 (February 1988).
6. J. H. Field, "Two Novel Special Relativistic Effects: Space Dilation and Time Contraction," *AJP* **68**, 367 (2000).
7. Robert W. Flynn, "The Relativistic Velocity Addition Formula," *TPT* **29**, 524 (November 1991).
8. John R. Graham, "Special relativity and length contraction," *AJP* **63**, 637 (July 1995).
9. Carlton A. Lane, "Space Travelers, Beware!," *TPT* **30**, 397 (October 1992).
10. David Merman, "The amazing many-colored relativity engine," *AJP* **58**, 601 (July 1988).
11. Cory J. Naddy, Scott C. Dudley, and Ryan K. Haaland, "Projectile Motion in Special Relativity," *TPT* **38** (January 2000).
12. C. Peters, "An alternate derivation of relativistic momentum," *AJP* **54**, 804 (September 1986).
13. Fritz Rohrlich, "An elementary derivation of $E = mc^2$," *AJP* **58**, 348 (1990).
14. R. Sandin, "In defense of relativistic mass," *AJP* **59**, 1032 (November 1991).
15. P. Sastry, "Is length contraction really paradoxical?," *AJP* **55**, 943 (1987).
16. Reinhard A. Schumacher, "Special relativity and the Michelson-Morley interferometer," *AJP* **82**, 609 (July 1994).

17. Edwin F. Taylor, "Why does nothing move faster than light? Because ahead is ahead!," *AJP* **58**, 889 (September 1990).
18. Volker Thomsen, "Signals from Communications Satellites," *TPT* **34**, 218 (April 1996).
19. David Z. Albert, "Bohm's Alternative to Quantum Mechanics," *SA* **270**, 58 (May 1994).
20. Francois Bardou, "Transition between particle behavior and wave behavior," *AJP* **59**, 458 (May 1991).
21. David G. Cassidy, "Heisenberg, Uncertainty and the Quantum Revolution," *SA* **266**, 106 (May 1992).
22. Robert Deltete and Reed Guy, "Einstein's opposition to the quantum theory," *AJP* **58**, 673 (July 1990).
23. Berthold-Georg Englert, Marlan O. Sailly, and Kerbert Walther, "The Duality of Matter and Light," *SA* **271**, 86 (December 1994).
24. Art Hobson, "Teaching Quantum Theory in the Introductory Course," *TPT* **34**, 202 (April 1996).
25. Richard Kidd, James Ardini, and Anatol Anton, "Evolution of the modern photon," *AJP* **57**, 27 (January 1989).
26. Jesusa Valdez Kinderman, "Investigating the Compton Effect with a Spreadsheet," *TPT* **30**, 426 (October 1992).
27. Abner Shimony, "The Reality of the Quantum World," *SA* **258**, 46 (January 1988).
28. A. Tonomura, J. Endo, T. Matsuda, T. Kawasaki, and H. Ezawa, "Demonstration of single electron buildup of an interference pattern," *AJP* **57**, 117 (February 1989).
29. M. P. Fewell, "The atomic nuclide with the highest mean binding energy," *AJP* **63**, 653 (July 1995).
30. Stig Steenstrup and Leif Gerward, "Becquerel's Discovery of Radioactivity - A Centenary," *TPT* **34**, 286 (May 1996).
31. Stephen G. Brush, "How Cosmology Became a Science," *SA* **267**, 62 (1992).
32. Wendy L. Freedman, "The Expansion Rate and Size of the Universe," *SA* **267**, 54 (November 1992).
33. Samuel Gulkis, Philip M. Lubin, Stephen S. Meyer, and Robert F. Silverberg, "The Cosmic Background Explorer," *SA* **262**, 132 (January 1990).
34. Jonathan J. Halliwell, "Quantum Cosmology and the Creation of the Universe," *SA* **265**, 76 (December 1991).
35. David Lindley, Edward W. Kolb, and David N. Schramm, eds., *Cosmology and Particle Physics,* AAPT, 1991.
36. David Lindley, Edward W. Kolb, and David N. Schramm, "Resource Letter CPP-1: Cosmology and particle physics," *AJP* **58**, 492 (June 1988).
37. Thomas A. Roman, "General Relativity, black holes, and cosmology: A course for nonscientists," *AJP* **54**, 144 (February 1986).

38. Richard Wilson, "Resource Letter EIRLD-1: Effects of Ionizing Radiation at Low Doses," *AJP* **67**, 372 (May 1999).

Demonstrations — Audio/Visual

American Association of Physics Teachers (AAPT) / ZTEK

- for the AAPT, see address information on page 31 and see the AAPT store site at http://www.aapt.org/store/. The store contains numerous books, computer programs for both Macintosh and PC, VHS tapes, DVDs, and more. The products are offered with discounts for members.

- for ZTEK, see http://www.ztek.com/physics/physics.html. You'll find video resources on DVD, CD-ROM ,and the videodisk format. Titles include:

Color Images of Physical Phenomena CD-ROM includes lessons covering 28 physics concepts along with an indexed database

Frames of Reference DVD

Physics Single Concepts Collections (DVD, videotape or videodisks with audio tracks and Instructor's Guides)

Collection 1 contains 35 segments from the Project Physics film loops that cover motion, modern physics, momentum, energy, waves, and collisions.

Collection 2 contains 38 segments from the Ealing film loops that cover mechanics, collisions, periodic motion, gases, light, and electricity and magnetism.

Miller Collection is a three video tape series (or video disc) created from 19 silent, short films covering mechanics, waves, atomic and nuclear physics.

Physics: Cinema Classics (videodisks and instructor's CD-ROM)
more than 245 classic video and still photo presentations of physics principles that cover mechanics, thermodynamics, waves, electricity and magnetism, light & optics, and modern physics, more information can be found at the following website: http://www.ztek.com

Physics Views: What If... DVD demonstrating 16 real-world events that demonstrate physics phenomena.

Skylab Video (30 min. video or DVD, 1975)
is a collection of 12 NASA silent film segments of astronauts demonstrating physics principles and phenomena in the Skylab facility.

Visual Quantum Mechanics CD-ROM contains 5 units on a single CD-ROM that may be used on either Macintosh or PC computer. The units include: solids & light, waves of matter, potential energy diagrams, luminescence, and exploring the very small (PC only).

Films for the Humanities and Sciences

PO Box 2053, Princeton, New Jersey 08543-2053
Phone: (800) 257-5126 or (609) 275-1400
web - http://www.films.com
e-mail: custserv@films.com

There are 277 items in the physics category, which are listed below. The website offers a detailed description of each title.

A Glorious Accident: Understanding Our Place in the Cosmic Puzzle
A Life of Time: Physics and Chronology
About Time
Acceleration
Albert Einstein
Alternative Power Sources and Renewable Energy
Atmospheric Pressure
Atoms and Molecules
Between Order and Disorder
Bipolarity
Bottling the Sun: The Quest for Nuclear Fusion
Capacity of a Condenser
Celestial Mechanics: The Distance to the Moon
Center of Mass Motion
Charging and Discharging
Charging by Induction
Chemical, Electrical, and Nuclear Energy
Chernobyl: Ten Days for Disaster
Chernobyl: The Dead Zone—on CD-ROM
Chernobyl: The Taste of Wormwood
Circular and Rotational Motion
Circular Motion
Climate Out of Control
Climbing Mount Improbable
Color
Commercial Generation and Transmission of Electricity
Conductors and Insulators
Conductors and Insulators
Conservation and Energy Alternatives: Powering the Future
Convex and Concave Lenses
Cp/Cv for Helium, Nitrogen, and Carbon Dioxide
Creating Magnetic Fields
Creating Magnetic Force
Current Electricity
Designed and Designoid Objects
Direct Current and Alternating Current
Domain Theory
Double Aperture Interference
Earth's Magnetic Field
Electric Current
Electric Fields Produced by a Van de Graaff Generator
Electrical Energy
Electrical Energy from Fission
Electricity
Electricity and Magnetism
Electricity in the Home
Electricity: Dry Batteries and Light Bulbs
Electromagnetic Induction

Electromagnetism
Electron Diffraction
Electron Waves Unveil the Microcosmos
Electronic Warfare: Microwaves and Weaponry
Electrons at Play: A Century of Electrifying Discoveries
Electrostatics 1
Electrostatics 2
Empowering the World: Technologies for a Sustainable Future
Energy
Energy Alternatives: Fusion
Energy Alternatives: Solar
Energy and Force: Part 1: Inertia, mass, speed, and acceleration. (25 min.)
Energy and Force: Part 2: Gravity, weight vs. mass, work, kinetic energy, and potential energy. (25 min.)
Energy from the Nucleus
Energy Today and Tomorrow
Energy: Nature's Power Source
Equilibrium of Forces
Eureka!
Experiments on the Doppler Effect
Exploring the Ionosphere
F = ma: Physical Science Concepts
Falling Motion
Fluid Pressure
Force
Force and Motion—on CD-ROM
Fresnel Lenses
Friction, Work, and Energy
From Particles to Waves: Electrons and Quantum Physics
Gathering Light
Gravitation
Green Energy
Ground Zero Plus Fifty: A Half-Century of Nuclear Experimentation and Devastation
Recommended: Booklist
Growing Up in the Universe
Half Lives: History of the Nuclear Age
Recommended: Video Librarian, Booklist
Harmful Effects of Electromagnetism
Harmonic Motion and Waves
Heat
Heat and Temperature
Hunt for the Elusive Neutrino
Inertia
Influence of a Force
Inside a Nuclear Plant
Interactive Physics—on CD-ROM
Introduction to Magnets
Introduction to Relative Motion
Introductory Concepts in Physics
Investigations in Physics: Experiments and Observations
Ionization and Excitation Potential
Length
Lenses
Life in the Field
Light and Shadow
Linear Momentum and Newton's Laws of Motion
Magnetic Fields
Magnetic Fields in Space: The Northern Lights

Magnetic Force
Magnetism and Electricity: Practical Applications
Magnetism and Electron Flow
Magnetism and Static Electricity
Mass
Matter and the Universe
Matter Waves
Measure for Measure
Mechanical, Thermal, and Light Energy
Mechanics in Action
Medical Applications of Electromagnetism
Meltdown
Millikan's Oil-Drop Experiment
Mirror and Image
Modeling Circular Motion
Modeling Satellite Motion
Molecular Machines Go to Work
Momentum
Motion
Motion of Bodies and Mechanical Energy
Movement and the Center of Gravity
Natural Transmutations
New Energy Sources
Newton's Revolution: Understanding Motion
Niels Bohr
Nuclear By-products
Nuclear Energy
Nuclear Materials: Russia's New Export
Nuclear Physics
Nuclear Power Plant Safety: What's the Problem?
Nuclear Technology
Nuclear Waste Disposal: Russia's Deep Secret
Numbers, Units, Scalars, and Vectors
One-Dimensional Kinematics
Optics
Oscillations and Waves—on CD-ROM
Our Hiroshima
Particle Physics
Patterns of Electromagnetic Fields
Photonics: The Revolution in Communications
Photons
Physical Processes—on CD-ROM
Physics in Action
Physics: Introductory Concepts
Physics: The Standard Deviants® Core Curriculum
Physics: The Standard Deviants® Core Curriculum - On DVD
Polarized Light
Potential Difference
Potential Energy and Kinetic Energy
Pressure
Projectile Motion
Properties of Becquerel Rays
pV Isotherms of CO_2: 1
pV Isotherms of CO_2: 2
Quarks and the Universe: Murray Gell-Mann
Radars
Radiation: Origins and Controls
Radiation: Types and Effects

Radioactive Waste Disposal: The 10,000-Year Test
Radioactivity
Radioactivity: Enemy or Friend?
Reflection of Light
Refraction of Light
Relative Circular Motion
Resistance
Richard Feynman: Take the World from Another Point of View
Riding on Air: The Principles of Flight
Robert Oppenheimer
Scientific Inquiry and Everyday Life: Steven Weinberg
Simple Machines
Simulated Intelligence
Smaller Than the Smallest
Spectra
Speed
Standing Waves
Structure of the Atom
Surface Interference Patterns
Sweet Disaster
Temperature
Temperature and Kinetic Theory
The Atom Revealed
The Bohr Model
The Brain Machine
The Conduction of Heat
The Convection of Heat
The Determination of a Radioactive Half-Life
The Determination of Boltzmann's Constant
The Determination of the Newtonian Constant of Gravitation
The Determination of the Velocity of Light
The Determination of the Velocity of Radio Waves
The Diffraction of Light
The Discovery of Radioactivity
The Earliest Models
The Effect of Pressure on the Thermal Conductivity of a Gas
The Electromagnetic Model
The Electromagnetic Spectrum
The Forces of Nature
The Generation of Current
The Generation of Electricity
The Genesis of Purpose
The Image
The Laws of Motion
The Laws of Motion Applied
The Life of Edward Teller
The Living Cell
The Magnet as Compass Needle
The Math Life
The Mother of All Collisions: The Dynamics of Impacts
The Motor Principle
The Nature of Waves
The Origin of Life
The Particle Model
The Physics and Physiology of Sports
The Physics of Amusement Park Rides
The Power of Sound
The Propagation of Waves

The Properties of Light
The Quantum Idea
The Relativity of Motion
The Rutherford Model
The Rutherford Scattering of Alpha Particles
The Search for Reality: The Story of Quantum Mechanics
The Solar Furnace
The Subject of Matter
The Theories of Physics
The Thermal Expansion of Metals
The Ultra-Microscopic World
The Ultraviolet Garden
The Uncertainty Principle: Making of an American Scientist
The Unification Theory
The Uses of Electricity: The Good, the Bad, and the Indefensible
The Wave Model
The Wave-Mechanical Model
The World of Absolute Zero
Thermodynamics
Thomas Alva Edison
Time
Understanding Electricity
Understanding Energy
Understanding Magnetism
Understanding Time
Understanding Uncertainty
Uses of the Electromagnet
Voltage
Waking Up in the Universe
Water-Lenses
Wave-Particle Duality
Waves and Sound
Waves and Vibrations: Information and Energy
Weight
Why Physics?
Why Planes Fly and Other Things

Jefferson Lab Science Series

Physics-related videotapes are available for two-week loan from the Thomas Jefferson National Accelerator Facility. The level of the material is pre-college, but may be of interest. The length of the tapes is probably too long for the lecture, but the tapes may be copied for students to look at in the library or A/V lab. To borrow one of the following tapes, contact the Education Office at Jefferson Lab at (757) 269 - 7564 or via e-mail at svideo@jlab.org.

Their website is: http://education.jlab.org/scienceseries/physics.html

Physics in Everyday Life (57 min., 1992)
What is the speed of light, anyway? (78 min, 1992)
Microscopes and Telescopes - How does the world work? (66 min, 1992)
Lasers - tunable and electron lasers (60 min., 1992)
Demonstrating the Laws of Physics - uses everyday objects to illustrate physics principles (60 min., 1993)

Neutrinos: Much Ado About (Almost) Nothing (60 min., 1993)
The world in a glass of soda pop (55 min, 1994)
Becoming Enlightened about Light (60 min, 1994)
Mathematics of Fluid Motion (50 min., 1996)
The Big Bang (50 min., 1996)
What goes up... (44 min, 1997)
Physics of Music (60 min, 1997)
Physics IQ Test (65 min, 1997)
Adventures in Science (56 min, 1997)
Clocks and Timekeeping (57 min, 1998)
Holograms (47 min, 1999)
Science Entertainment (57 min, 1999)
Physics Circus (52 min, 1999)
Radiation: What is it and how can it affect me? (55 min, 2000)
The Physics of Baseball (52 min, 2002)
Einstein's Unfinished Symphony (60 min, 2002)
Crazy Ideas in Science (59 min, 2002)
Particle Odyssey (55 min, 2002)

Media Design Associates

Box 3189, Boulder, Colorado 80307-3190 Phone: (800) 228-8854
web - http://indra.com/mediades/index.html

Demonstrations of Physics: Motion (video, segments of varying length) - titles: *The Ape in the Tree* (3:13), *Newton's First Law* (5:40), *Newton's Second Law* (6:31), and *Uniform Circular Motion* (5:57)

Demonstrations of Physics: Energy and Momentum (video, segments of varying length) - titles: *Air Track Collisions* (7:24), *The Dynamics of a Karate Punch* (4:55), *Conservation of Angular Momentum* (5:52), and *Why a Spinning Top Doesn't Fall* (4:24)

Demonstrations of Physics: Liquids and Gases (video, segments of varying length) - titles: *Why Divers Exhale While Surfacing* (2:25), *Archimedes' Principle for Gases* (3:18), *The Cartesian Diver* (6:27), and *Bernoulli's Equation and Streamlines* (5:33)

Demonstrations of Physics: Thermal Effects (video, segments of varying length) - titles: *Specific Heat* (2:25), *Expansion Due to Heating* (7:26), and *Heat of Fusion* (6:14)

Demonstrations of Physics: Waves (video, segments of varying length) - titles: *Wave Characteristics* (7:50), *Measuring the Speed of Sound* (7:48), *Chladni Plates* (4:46), and *Resonance* (4:32)

Demonstrations of Physics: Electricity (video, segments of varying length) - titles: *Static Electricity* (7:12), *Electrostatic Generators* (10:54), *Series and Parallel Circuits* (5:16), *Magnetic Fields* (6:14), and *Electromagnetic Induction* (7:33)

Demonstrations of Physics: Light (video, segments of varying length) - titles: *Reflection* (6:34), *Refraction* (8:07), *Spectra* (6:02), *The Dual Nature of Light* (8:16), and *Viewing a Hologram* (3:36)

Matter (videodisk or VHS) topics include: solids, liquids, gases, molecular structure, surface tension, atmospheric pressure, specific heat, density, phase changes, melting and boiling points, thermal conductivity, freezing and the First Law of Thermodynamics.

Physics Curriculum and Instruction

22585 Woodhill Drive, Lakeville, Minnesota 55044
Phone (612) 461-3470 or Fax (612) 461-3467
web - http://www.physicscurriculum.com
The following videos are available in both VHS and DVD formats.

Physics Demonstrations in Mechanics (video series, 3 min. each segment) - uniform and accelerated motion, gravitational acceleration, graphical analysis of motion, vector addition, projectile motion, circular motion, velocity & acceleration vectors, projectile motion, Newton's laws of motion, terminal velocity, work & energy, conservation of energy, center-of-mass motion, conservation of momentum, impulse & momentum, collisions, rotational dynamics, angular momentum conservation, and Newton's law of universal gravitation

Physics Demonstrations in Heat (video series, 3 min. each segment) - phase changes, phase change expansion, specific heat, thermal conduction, thermal convection, heat transfer mechanisms, thermal radiation, mechanical equivalent of heat, kinetic model, and entropy

Physics Demonstrations in Sound and Waves (video series, 3 min. each segment) - wavelength & frequency, mechanical resonance, standing waves, interference, longitudinal waves, longitudinal standing waves, nature of sound, propagation of sound, transmission of sound, refraction of sound, Doppler effect, standing sound waves, and the superposition principle

Physics Demonstrations in Electricity and Magnetism (video series, 3 min. segments) - electrostatics, isolation of charges, electric fields, temperature & resistance, magnetic fields, electricity & magnetism, electromagnetic effects, induction applications, and eddy currents

Physics Demonstrations in Light (video series, 3 min. segments) - propagations of light, the visible & infrared spectrum, interference & interferometers, and thin film interference

Also available are: *Physics of Space Flight Series: Parts I, II, III* and *Seeing the Physical World with High-Speed Cameras.*

Other Audio/Video Resources

The Mechanical Universe (videotapes or videodisk set)
famous series of fifty-two 30 min. segments covering an entire freshman level physics course that was produced at CalTech; some segments contain calculus, nice use of graphics and historical perspectives, contact: Annenberg CPB Project at PO Box 2345, South Burlington, VT 05407-2345, phone (800)-LEARNER, or visit the following website:

http://www.its.caltech.edu/~tmu/

The Physics of Sports (videodisk)
available from Videodiscovery at (800)-548-3472

The Powers of Ten (video)
video zooms from the subatomic realm to the size of the universe, contact the W. H. Freeman company at (800) 877-5351 or it may be ordered through a book store with the following: ISBN 0-7167-5029-5. It is also available from http://www.amazon.com.

Demonstrations: Computer Software

Computers have become increasingly popular in physics education both inside and outside the classroom. Computers are being used for simulations, playing video clips and analyzing them, data analysis and graphing, interactive tutorials, and for practice tests. Students in some classrooms have networked laptop computers to provide an interactive lecture environment. With the continuing advancement of the multimedia capabilities of computers (along with increases in memory storage and speed), their usage in education will probably continue to increase. As with any educational tool, however, the instructor needs to explain how to use the software and what the goals are in its use. Successful education is still dependent on the quality of the interaction between the instructor and the students. Today's computer software can make physics fun and interesting to learn. *One important caution:* don't let students get caught up in the software that models or illustrates something physical and expect intuition of the physical world to develop. Students still need hands-on, trial and error experiments to get that.

Print, Audio/Visual, and Computer Resources

> **Available from your Wiley Representative or by calling (800) 594 – 5396.**
> See also - http://he-cda.wiley.com/WileyCDA/
>
> ***Instructor's Resource CD-ROM*** - This supplement contains the entire *Instructor's Solutions Manual* (both MS Word and PDF file formats), the *Computerized Test Bank* of 2200 questions in multiple-choice format (for both PC and Macintosh platforms) with full editing functionality, Multiple-choice format for approximately 1200 even-numbered homework problems in the text, all text illustrations (for classroom projection, printing, or secure web posting), and simulations for classroom demonstrations.
>
> ***Cutnell Multimedia*** - This CD-ROM supplement contains the complete *Physics, 6e* text, the *Student Study Guide*, the *Student Solutions Manual*, the *Interactive LearningWare*, and numerous simulations, all connected with extensive hyperlinking.
>
> ***Wiley Physics Simulations*** - This CD-ROM contains 50 physics simulations programmed in Java. These can be used for lecture demonstrations or online student assignments. Easy interface for adjustment of input data and provides both graphical and numerical output.

The resources given below are in addition to those available from the internet and mentioned in Chapter 2.

- ***Beam Two*** - advanced ray tracing software for Macintosh and Windows is available from Stellar Software, PO Box 10183, Berkeley, CA 94709, or phone (510) 845 - 8405, e-mail ssw@dnai.com, web http://www.stellarsoftware.com

- ***Cartoon Guide to Physics*** - an inexpensive CD-ROM based on the book by Larry Gornick and Art Hoffman. Although the user is guided by a cartoon character, the physics is treated seriously as the user is guided through various aspects of mechanics. The disk also includes biographical information about well-known physicists. The CD-ROM is produced by HarperCollins Interactive Publishers, 10 East 53 Street, New York, 10022, or call (800) 424 - 6234.

- ***OnScreen Particle Physics*** - an award-winning Macintosh/Windows simulation program for particle-chamber demonstrations is available from OnScreen Science, 46 Wallace Street, Sommerville, MA 02144-1807, email info@onscreen-sci.com, or see website http://www.onscreen-sci.com/

Interactive Physics - *Interactive Physics Player Workbook* by Cindy Schwarz with CD-ROM for either Macintosh or Windows (ISBN 0130671088) is a text-independent workbook containing approximately 200 examples and problems organized topically, each simulated by the Interactive Physics Player. The book contains tutorial "what if" questions and explanation and the software. Provides graphic interactive simulation by allowing students to change parameters and observe results. For introductory algebra or calculus-based physics courses.

Physics Academic Software
Box 8202, NCSU, Raleigh, NC 27695-8202
Phone (800) 955-8275 or fax (919) 515 2682 or web - http://webassign.net/pasnew/

Physics Academic Software (PAS) publishes software in cooperation with the American Institute of Physics (AIP), The American Physical Society (APS), and the American Association of Physics Teachers (AAPT). PAS reviews, selects, and publishes high-quality software suitable for use in high school, undergraduate, and graduate education in physics. All software is peer-reviewed for excellence in pedagogical or research value and tested for accuracy, compatibility, and ease of use. Packages include detailed documentation for users and instructors.

Titles include: *Atomic Scattering* (Macintosh), *Atoms in Motion* (Windows/Mac), *Audioscope* (Win), PEARLS *3.0 and 4.0* (complete curriculum package for Mac/Windows), *Chart of the Nucleides: A Tutorial* (Mac), *Conceptual Kinematics* (Mac/Dos), *DC Circuits* (DOS), *Dipole Magnets* (Win/Mac), *Electric Field Hockey 3.0* (Mac/DOS/Win), *EM Field 6* (Mac/DOS/Win), *Force and Motion Microworld* (DOS), *Forces* (Win/Mac), *Fourier Series in Mathematical Physics* (DOS), *Freebody* (Win/DOS/Mac), *Geometric Optics* (Mac/Win), *Graphical Schrödinger's Equation* (Win), *Lighting Up Circuits* (Mac), *Mechanics in Motion* (Win), *Motion in Electromagnetic Fields* (Mac/Win), *Objects in Motion* (DOS/Mac), *Optics Phenomena* (Win), *Photoelectric Tutor* (Win/DOS/Mac), *Physics Demonstrations* (DOS), *Physics of Oscillations* (Win/DOS), *Physics Plot* (Win), *PhysWiz* (Win), *Planets and Satellites* (Win), *Quadrupole Magnets* (Mac/Win), *Ray* (Mac), *RelLab* (relativity, Mac), *Solid State Physics* (DOS), *Spacetime* (DOS/Mac), *CUPS Utilities* (DOS), *MUPPET Utilities* (DOS), *Vectors* (Win/Mac), *VideoGraph* (Mac), *Virtual E Field Lab* (Win), *Wave Interference* (DOS), *WaveMaker* (Mac)

Physics Tutor
web - http://www.highergrades.com/physicstutor/home.html

This is interactive educational software geared toward high school and freshman college physics.

CHAPTER 4

Usage of the Text and Supplementary Materials

The 32 chapters in *Physics, 6e* by John D. Cutnell and Kenneth W. Johnson are arranged in the fairly standard sequence for introductory physics courses: mechanics (chapters 1 through 11), thermal physics (chapters 12 through 15), wave motion (chapters 16 and 17), electricity and magnetism (chapters 18 though 23), light and optics (chapters 24 through 27) and modern physics (chapters 28 through 32). New material and features have been added :

• ***Check Your Understanding*** - These are short, one-question conceptual self-quizzes that are intended to help the students assess their understanding as they proceed through the chapter. There are 3 or 4 per chapter and answers are provided. This is a very effective feature for getting students to become active readers.

• The redesigned, expanded ***Chapter Summary*** emphasizes key concepts and provides a clear road map to all the important pedagogical features - both in print and online.

• The highly interactive **Student Web Site** has been enhanced with the addition of two new features: ***Self-Assessment Tests*** and ***Simulation Exercises***. These new features, along with the existing Interactive LearningWare problem-solving tutorials, are highlighted in the text with clear explanations of the student benefits. Altogether, the Student Web Site provides students not only with a rich set of study aids but also with interactive material that allows them to work with the material in a way that is not possible with print alone.

It would be difficult and unwise to try to cover *all* of the text material in a one-year course. This chapter is meant to serve as an aid to the instructor in gaining a broad overview in the planning the use of the text and its supplements in the one-year course. In Chapter 5, you are invited to plan the course in detail using the information (teaching objectives, lists of transparencies, lecture demonstrations, and suggested laboratory experiments) provided for each chapter and the space provided for each section in the text.

The Semester System

For schools on the semester system, mechanics, thermal physics, and wave motion are usually covered during the first semester; and the remaining material is presented during the second semester. The tables on the following page contain a suggested schedule for a two-semester course. In preparing the tables, lecture time was assumed for three 50-minute or two 75-minute sessions per week for fourteen weeks each semester. This results in forty-one meetings per semester for 50-minute classes and in twenty-eight meetings for 75-minute classes. In preparing the schedules, it was also assumed that, the first session would be an introduction to the course and the final class would be a review session. Time for exams or quizzes has not been included in the tables because the amount of time required for them will vary depending on the number and length of the tests. The exact number of meetings will also vary depending on the school calendar. The suggested schedules should be viewed as the average amount of time required for presentation of the material from *all of the non-starred sections* of the text.

Another consideration for lecture planning is the allowance of a certain amount of time at the beginning of each lecture for questions, comments, announcements, and a short review of material from the previous lecture. The review helps provide a seamless presentation of the material and prepare everyone to receive the new material.

Suggested Schedule for a Two Semester Course

	50 minutes / 3 per week			75 minutes / 2 per week		
	Chapter	Number of Lectures	Cumulative Lectures	Chapter	Number of Lectures	Cumulative Lectures
first semester	1	2.0	2.0	1	1.3	1.3
	2	2.5	4.5	2	1.7	3.0
	3	2.0	6.5	3	1.3	4.3
	4	3.5	10.0	4	2.3	6.6
	5	1.8	11.8	5	1.2	7.8
	6	2.4	14.2	6	1.6	9.4
	7	1.8	16.0	7	1.2	10.6
	8	1.5	17.5	8	1.0	11.6
	9	2.5	20.0	9	1.7	13.3
	10	2.2	22.2	10	1.5	14.8
	11	3.0	25.2	11	2.0	16.8
	12	2.8	28.0	12	1.8	18.6
	13	1.4	29.4	13	1	19.6
	14	1.5	30.9	14	1	20.6
	15	3.1	34.0	15	2.1	22.7
	16	3.0	37.0	16	2.0	24.7
	17	2.0	39.0	17	1.3	26.0

	50 minutes / 3 per week			75 minutes / 2 per week		
	Chapter	Number of Lectures	Cumulative Lectures	Chapter	Number of Lectures	Cumulative Lectures
second semester	18	3.0	3.0	18	2.0	2.0
	19	3.0	6.0	19	2.0	4.0
	20	4.0	10.0	20	2.5	6.5
	21	3.5	13.5	21	2.3	8.8
	22	3.5	17.0	22	2.4	11.2
	23	2.6	19.6	23	1.6	12.8
	24	2.0	21.6	24	1.3	14.1
	25	2.0	23.6	25	1.4	15.5
	26	4.4	28.0	26	3.0	18.5
	27	3.0	31.0	27	2.0	20.5
	28	1.5	32.5	28	1.0	21.5
	29	1.5	34.0	29	1.0	22.5
	30	2.0	36.0	30	1.5	24.0
	31	1.5	37.5	31	1.0	25.0
	32	1.5	39.0	32	1.0	26.0

The Quarter System

For schools on the quarter system, mechanics is usually covered during the first quarter; and the remaining material is presented in one of two ways during the second and third quarters. The tables below contain two suggested schedules for the three-quarter course. In the first schedule, the material is presented in the sequence as the text, but this results in splitting the electricity and magnetism portion between quarters. The second schedule does not split any major subject area. Instead, light and optics are presented directly after the chapters on wave motion in the second quarter. Then, electricity and magnetism and the chapters on modern physics are presented in the third quarter. In preparing the tables, lecture time was assumed for three 50-minute or two 75-minute sessions per week for ten weeks each quarter. This results in thirty meetings per semester for 50-minute classes and in twenty meetings for 75-minute classes. In preparing the schedules, it was also assumed that, the first session would be an introduction to the course and the final class would be a review session. Time for exams or quizzes *has been included* in these tables (2 or 3 mid-term exams) even though the amount of time required for them will vary depending on the number and length of the tests. The exact number of meetings will also vary depending on the school calendar. The suggested schedules should be viewed as the average amount of time required for presentation of the material from *all of the non-starred sections* of the text.

Suggested Schedule for a Three Quarter Course
(with Electricity & Magnetism Split Between Quarters)

first quarter

	50 minutes / 3 per week			75 minutes / 2 per week	
Chapter	Number of Lectures	Cumulative Lectures	Chapter	Number of Lectures	Cumulative Lectures
1	2.1	2.1	1	1.4	1.4
2	2.4	4.5	2	1.5	2.9
3	1.8	6.3	3	1.2	4.1
4	3.6	9.9	4	2.5	6.6
5	1.6	11.5	5	1.1	7.7
6	2.3	13.8	6	1.5	9.2
7	1.7	15.5	7	1.4	10.6
8	1.4	16.9	8	1.0	11.6
9	2.6	19.5	9	1.7	13.3
10	2.3	21.8	10	1.6	14.9
11	3.2	25.0	11	2.1	17.0

second quarter

	50 minutes / 3 per week			75 minutes / 2 per week	
Chapter	Number of Lectures	Cumulative Lectures	Chapter	Number of Lectures	Cumulative Lectures
12	3.0	3.0	12	2.1	2.1
13	1.7	4.7	13	1.1	3.2
14	1.7	6.4	14	1.1	4.3
15	3.4	9.8	15	2.3	6.6
16	2.9	12.7	16	2.1	8.7
17	2.1	14.8	17	1.4	10.1
18	3.1	17.9	18	2.1	12.2
19	2.8	20.7	19	1.9	14.1
20	4.3	25.0	20	2.9	17.0

Usage of the Text and Supplementary Materials

third quarter

	50 minutes / 3 per week			75 minutes / 2 per week	
Chapter	Number of Lectures	Cumulative Lectures	Chapter	Number of Lectures	Cumulative Lectures
21	3.3	3.3	21	2.1	2.1
22	3.2	6.5	22	2.1	4.2
23	2.1	8.6	23	1.4	5.6
24	1.5	10.1	24	1.0	6.6
25	1.6	11.7	25	1.1	7.7
26	3.8	15.5	26	2.5	10.2
27	2.8	18.3	27	1.9	12.1
28	1.3	19.6	28	1.0	13.1
29	1.0	20.6	29	0.7	13.8
30	1.8	22.4	30	1.2	15.0
31	1.3	23.7	31	1.0	16.0
32	1.3	25.0	32	1.0	17.0

Suggested Schedule for a Three Quarter Course
(with No Split Between Any Main Sections)

first quarter

	50 minutes / 3 per week			75 minutes / 2 per week	
Chapter	Number of Lectures	Cumulative Lectures	Chapter	Number of Lectures	Cumulative Lectures
1	2.1	2.1	1	1.4	1.4
2	2.4	4.5	2	1.5	2.9
3	1.8	6.3	3	1.2	4.1
4	3.6	9.9	4	2.5	6.6
5	1.6	11.5	5	1.1	7.7
6	2.3	13.8	6	1.5	9.2
7	1.7	15.5	7	1.4	10.6
8	1.4	16.9	8	1.0	11.6
9	2.6	19.5	9	1.7	13.3
10	2.3	21.8	10	1.6	14.9
11	3.2	25.0	11	2.1	17.0

second quarter

	50 minutes / 3 per week			75 minutes / 2 per week	
Chapter	Number of Lectures	Cumulative Lectures	Chapter	Number of Lectures	Cumulative Lectures
12	2.9	2.9	12	1.9	1.9
13	1.5	4.4	13	1.0	2.9
14	1.6	6.0	14	1.1	4.0
15	3.2	9.2	15	2.2	6.2
16	2.9	12.1	16	2.0	8.2
17	1.9	14.0	17	1.3	9.5
24	1.9	15.9	24	1.3	10.8
25	1.9	17.8	25	1.3	12.1
26	4.3	22.1	26	2.8	14.9
27	2.9	25.0	27	2.1	17.0

third quarter

	50 minutes / 3 per week			75 minutes / 2 per week	
Chapter	Number of Lectures	Cumulative Lectures	Chapter	Number of Lectures	Cumulative Lectures
18	2.7	2.7	18	1.8	1.8
19	2.5	5.2	19	1.7	3.5
20	3.8	9.0	20	2.5	6.0
21	3.3	12.3	21	2.2	8.2
22	3.3	15.6	22	2.2	10.4
23	2.2	17.8	23	1.5	11.9
28	1.4	19.2	28	1.1	13.0
29	1.1	20.3	29	0.8	13.8
30	1.9	22.2	30	1.2	15.0
31	1.4	23.6	31	1.0	16.0
32	1.4	25.0	32	1.0	17.0

Getting the Most from the Text Supplements

While the text is certainly self-contained for both study and teaching, both tasks can be made easier by using the text supplements. The schematic diagram below illustrates the overall structure of the course and the relationships between the instructor, students, the text, and the various supplements. In the sections that follow, each supplement is discussed and suggestions are made for its usage.

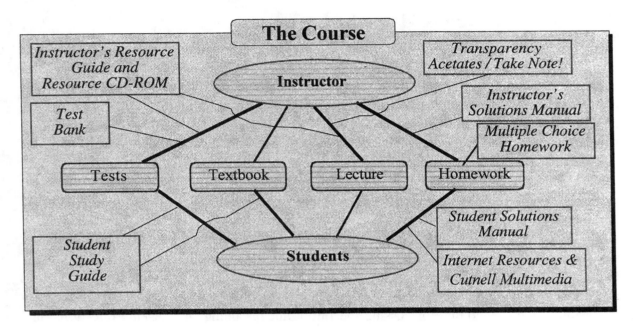

Student Study Guide

Just as this *Instructor's Resource Guide* is designed to be an aid to instructors in preparing, organizing, and managing the introductory physics course, the *Student Study Guide* (SSG) is the equivalent aid the student. The SSG is used in conjunction with the text, not in place of it. Students should carefully read the text and the corresponding material in the SSG that includes: chapter previews, lists of important terms, and topical discussions with particular focus on areas that students find troublesome. After completing this reading, the student is then invited to work through additional examples, practice problems, and end-of-chapter quizzes. For pre-med students, each chapter has material and problems that will help them prepare for the MCAT.

Students should be made aware of the existence of the study guide if it is not already suggested at the time of the textbook purchase. Tell the students what the guide is for and point out that it can be a good way to organize their study. In my classes, I point out that the purchase will help them *if* they use the guide as it is intended. Many students buy the guide and initially find it useful, but unfortunately, as the semester progresses and their work load increases, they begin to use it less often (primarily as a review for quizzes and the final). Other students have told me that they were able to use it throughout the semester and felt that it had improved their grade by as much as 5 %, which can be the difference between getting a *B* and getting an *A*.

Supplements: Student Solutions Manual

Carefully written solutions to approximately 600 of the 2400 homework problems in the text are provided for students. The problems have been selected from among the odd-numbered homework problems; and those that are included in the *Student Solutions Manual* are identified with the label **ssm** in the text. The solutions contain explicit reasoning steps in which the procedure for approaching the problem and solving it is discussed before any algebraic or numerical work is done.

The SSM can be of great benefit to students who want to sharpen their problem-solving skills and improve their conceptual understanding of physics. Students should practice problem-solving on their own by selecting a problem from the text with the SSM label, working out the problem, and then comparing their work with that in the SSM. Eventually, students will get the idea that the reasoning step is the necessary first step instead of simply looking for an equation. Encourage students to use the SSM either individually or in groups and to read the solution very carefully. Using the SSM in a group environment can be very rewarding if they spend time discussing the important reasoning step, asking questions, and understanding the problem, before moving on to the algebraic and numerical solutions. Of course, students will also find the SSM useful in gaining additional practice in working problems in particular subject areas before a quiz or the final exam.

Supplements: Instructor's Solutions Manual

The solutions manual contains detailed solutions to all 2400 homework problems as well as the conceptual questions located at the ends of the chapters in the text. The manual has been conveniently divided into two volumes. The first volume contains solutions for chapters 1 through 17; and the second volume contains those for chapters 18 through 32. Many of the solutions to the homework problems contain the explicit reasoning step that is so very important in helping students learn physics and problem-solving skills.

Instructors find this supplement incredibly valuable because of the great care that has gone into preparing each solution and into carefully rechecking every solution to ensure that the answers are indeed correct. Having detailed solutions available saves the time that would be required for an instructor or a teaching assistant to solve the problems. Also, many instructors photocopy the individual solutions for a given homework assignment and post them for the class after the homework has been collected. Students should be encouraged to check the posted solutions against their own work. Care should be taken at all times to prevent student access to the solutions manual. Also, if problem solutions have been posted in previous semesters, there is a good probability that students in the current class will have access to them. In preparing homework assignments that are to be graded, take into account problems assigned during past semesters, the problem solutions available in the SSM, and those available via the internet at the Cutnell & Johnson *Physics* website to prevent students having an unfair advantage over students that do not have the same access.

Test Bank & Computerized Test Bank

The *Test Bank* contains 2200 short answer conceptual questions and quantitative problems in a multiple-choice format. The questions and problems have been organized into their respective sections within each chapter. A *level of difficulty* icon has also been added for each problem.

The computerized version of the test bank is available for both Macintosh and Windows-based computers. The software is very easy to use and provides full-editing capability.

This is another labor saving device for instructors, especially if the software is used to format and prepare the exams. The large number of questions of varying difficulty provides a great deal of flexibility. The organization scheme makes exam preparation easier than ever because one can quickly find a question concerning a given topic. Each problem has five answers labeled a, b, c, d, and e from which to choose. The answers have been randomized; and care has been taken to ensure a near uniform distribution among the five choices. Instructors who prefer using written tests may still use the questions and problems in the test bank by simply omitting the answer choices.

Homework Disk

Software is available for both Macintosh and Windows-based computers for the preparation of multiple choice homework assignments from the nearly 1200 even-numbered homework problems at the ends of the chapters in the text. The entire listing of the problems and the multiple choices is included in the *Test Bank*.

Supplements: Transparency Acetates / *Take Note!*

Two hundred illustrations from the text are available for use in the classroom. These acetates are high-quality, four color illustrations that have been resized and edited for maximum effectiveness when projected.

The overhead projector is one of the most common tools used for classroom teaching. Instructors often prepare their own transparencies from their notes and include equations, definitions, examples, etc. Most of us though are not very artistic when it comes to preparing drawings of some physical situation we plan to discuss in class. Often, student understanding of the situation rests solely on the illustration and its quality. This is why the color transparencies from the text are so valuable. The figure numbers of the available transparency acetates are listed in Chapter 5 of this guide for each chapter of the text to aid in course planning.

Take Note! is a notebook format on which students can take notes next to 8.5 inch by 11 inch, grayscale versions of the illustrations from the text, including the set of 200 overhead transparencies. The hope is that students save time by not having to copy the illustration into their notebooks and that they will have more accurate notes to the discussion.

Supplements: Instructor's Resource CD

This CD-ROM is an essential tool for every instructor. It contains the Test Bank and Multiple-choice Homework set with full-editing capability, the Instructor's Solution Manual in both Microsoft Word and Adobe Acrobat (pdf) formats, all text illustrations, and simulations for the classroom. The CD is available from your Wiley representative. To determine your representative's contact information, visit the Wiley Higher Education website at http://he-cda.wiley.com/WileyCDA/

Supplements: Text Website ~ http://www.wiley.com/college/cutnell

As Chapter 2 indicated, the internet has exploded into every aspect of life and is changing the way in which people communicate with each other. This is no less true for education. The Cutnell website has been designed to bring numerous resources to both instructor and student that expand the capabilities of the printed text. The online features are described as follows:

Usage of the Text and Supplementary Materials

eGrade

Assign, deliver, grade, and route homework, quizzes, and exams-- automatically! Developed by Professor John Orr and the faculty and graduate students at the University of Nebraska - Lincoln, Wiley eGrade is web-based software that automates the assigning, delivering, grading, and routing of homework, quizzes, and exams. Wiley eGrade features state-of-the-art capabilities to support content across all academic disciplines. It provides immediate feedback to students on their work. eGrade is also the first homework management solution to manage a wide range of math-based questions required in technical disciplines.

WebCT

Wiley and WebCT have partnered to create a series of e-Packs that provide course content from the text specifically designed for use in WebCT's software. Wiley e-Packs are the ideal solution for faculty looking for an easy-to-implement course management option. They enable instructors to start teaching online without having to create a course from scratch. The e-Packs include:
* Fully customizable course materials specifically designed for use with WebCT
* Special features such as animations, sample syllabi, lecture notes, quiz and test banks, and glossaries
* The flexible, easy-to-use functionality of WebCT's course management software

WebAssign

WebAssign was developed by teachers for teachers. Create assignments from their ready-to-use database of textbook questions, or write and customize your own exercises. The database contains approximately 1550 end of chapter conceptual questions and problems. You have complete control over the homework your students receive, including due date, content, feedback, and question formats.
- Create, post, and review assignments 24 hours a day, 7 days a week
- Deliver, collect, grade, and record assignments instantly
- Offer more practice exercises, quizzes, and homework
- Randomize numerical values or phrases to create unique questions
- Assess student performance to keep abreast of individual progress
- Capture the attention of your online, distance learning students

Student Companion Website
- *Interactive LearningWare Problems* These are interactive calculational examples presented using the Macromedia Flash Player. Each example consists of five steps: (1) visualization, (2) reasoning, (3) identifying knowns and unknowns, (4) modeling the problem, and (5) numerical solution.
- *Solutions to selected end-of-chapter problems*
- *Integration of Concepts Essays*
- *Practice Quizzes* – each chapter, mid-term, and MCAT self-quizzes
- *Simulation Exercises* - 50 Shockwave simulations covering topics throughout the text.

- *Interactive Illustrations* These are Macromedia files that bring some of the text's illustrations to life.
- *WebLinks* The web has been searched for websites containing content illustrative of the material in each chapter of the text. These links are useful for both instructors and students. As for the links listed in Chapter Two of this book, the links on the website may become outdated as time passes. You can use one of the search engines to try and locate the same content elsewhere, should this occur.

CHAPTER 5

Lecture Planning & Notes

No matter what your lecture style is, there is no substitution for carefully planned and executed class time. For most instructors, though, there just isn't enough time to plan every minute of every lecture. This section of the Resource Guide is designed to be a tool for planning, organizing, and improving your college physics course.

Each chapter and all of the sections of each chapter of the text are listed as well as teaching objectives, available transparencies, laboratory experiments, and lecture demonstrations. The *Concepts at a Glance* feature helps students see that physics is an integrated body of knowledge.

Some Suggestions for Using These Planning Resources:

- Each section has a box for time management. Estimate the time, T_{est} you would like to spend on each topic in planning your lectures. After class, record the actual time spent, T_{act}, and use this information for planning when you have to teach the course again.

- Space has been provided for each topic. You can use this space in a variety of ways:
    1. Plan lecture demonstrations or use of A/V materials. After class, indicate the relative success or failure of the demonstration and ways to improve the presentation.
    2. Indicate transparencies to include in the lecture presentation.
    3. Record interesting questions asked by students about the topic. Take note of students' misconceptions.
    4. Copy equations for use in lecture. If you hand out an equations sheet for exams, equations can be written down for inclusion on the sheet.
    5. Write down ideas for exam questions.
    6. Note which topics relate directly to the experiments students will be doing in their laboratory sections.
    7. Indicate an interesting anecdote, a news story, a "real world" application, a historical account, or a joke that you want to include in discussing the topic.

- An additional area has been included at the end of each chapter for additional notes and ideas.

Laboratory Experiments

Experiments are listed for each chapter from:
Laboratory Manual to accompany Physics by Cutnell & Johnson
John Wiley & Sons, Inc. (2003)
Pasco Scientific, Inc.
ISBN: 0-471-22699-8

Annotated Instructor's Edition with CD Cutnell)
0-471-29750-X
Includes answers, notes, and sample data. CD includes data files.

PASCO Scientific is the leading seller of equipment for physics labs. Working with physics educators, they have a developed a new line of equipment that has revolutionized the way physics labs are taught. Wiley is partnering with PASCO to produce a comprehensive set of experiments that allow teachers to fully implement this new program. There are extensive cross-references to Physics by Cutnell and Johnson.

Features
- First comprehensive set of labs incorporating the new interfacing technology and the new lab pedagogy.
- PASCO equipment is very widely used in physics labs.
- Annotated Instructors Edition provides extensive support for teacher.
- Can be customized (WCS) to accommodate local variations.

Lecture Demonstrations

There are relatively few books available for lecture demonstrations for an introductory physics course. Chapter 3 includes coverage of many aspects of lecture demonstrations and has an extensive bibliography. Four books are referenced in listing demonstrations for each chapter of the text:

- **The Dick and Rae Physics Demo Book**
  by D. Rae Carpenter, Jr. and Richard B. Minnix (1993)
  Dick and Rae, Inc.
  Lexington, VA 24450-0304

- **Physics Demonstration Experiments at William Jewell College**
  by Wallace A. Hilton (1982)
  American Association of Physics Teachers
  One Physics Ellipse
  College Park, MD 20740-3845
  (301) 209-3300

Lecture Planning and Notes

- **A Demonstration Handbook for Physics**
  by G. D. Freier and F. J. Anderson (1981)
  American Association of Physics Teachers
  One Physics Ellipse
  College Park, MD 20740-3845
  (301) 209-3300

- **Physics Demonstration Experiments**
  by Harry F. Meiners, ed. (1970)
  This book is currently out of print, but you may try your library or get it through an interlibrary loan.

1 Introduction and Mathematical Concepts

Teaching Objectives
- Define physics and explain its role and scope. Show students why it's important to learn and understand physics and how physics can be useful in their everyday lives and future careers.
- Introduce the fundamental tools of physics that will be the basis for all further study: basic units, unit conversions, scalars and vectors, and vector mathematics.
- Review basic trigonometric, geometric, and algebraic relations.

List of Transparency Acetates

Figure	Use (✓)	Notes
1.14		
1.18		
1.21		

Lecture Demonstrations
Carpenter and Minnix: M-016, M-018, and M-034
Freier and Anderson: Ma1-3 and Mb-2,3
Hilton: M-1 and M-10
Meiners: 6-4.7, 6-4.8, and 6-4..9

Laboratory Exercises
No laboratory exercises are available for this material.

Section 1.1 The Nature of Physics

T_{est} = ____ m
T_{act} = ____ m

Section 1.2 Units

T_{est} = ____ m
T_{act} = ____ m

Section 1.3 The Role of Units in Problem Solving

T_{est} = ____ m
T_{act} = ____ m

Section 1.4 Trigonometry

T_{est} = ____ m
T_{act} = ____ m

Section 1.5 Scalars and Vectors

T_{est} = ____ m
T_{act} = ____ m

Section 1.6 Vector Addition and Subtraction

T_{est} = _____ m
T_{act} = _____ m

Section 1.7 The Components of a Vector

T_{est} = _____ m
T_{act} = _____ m

Section 1.8 Addition of Vectors by Means of Components

T_{est} = _____ m
T_{act} = _____ m

Section 1.9 Concepts & Calculations

T_{est} = _____ m
T_{act} = _____ m

*A*dditional Notes and Ideas   Chapter 1

Additional Notes and Ideas (*continued*)

2 Kinematics in One Dimension

Teaching Objectives
- Introduce displacement and discuss the difference between displacement and distance.
- Introduce speed and velocity. Discuss the difference between average speed and average velocity and between average and instantaneous velocity. Make sure students understand the notation used in the text.
- Introduce average and instantaneous acceleration and discuss their correct usage. Illustrate the difference between velocity and acceleration.
- Introduce and discuss the correct usage of the equations of kinematics for an object moving with a constant acceleration along a straight line. Provide several interesting examples, pointing out problem solving strategies.
- Describe free fall and introduce the acceleration due to gravity.

Concepts at a Glance

In Section 2.2, **displacement** (Section 2.1) and **time** (Section 1.2) are brought together to formulate the concept of **velocity**.

In Section 2.3, to formulate the concept of **acceleration**, the change in **velocity** (Section 2.2) is combined with the **time** (Section 1.2) required for the change to occur.

In Section 2.4, the **equations of kinematics for constant acceleration** are obtained by combining the concepts of **displacement** (Section 2.1), **velocity** (Section 2.2), and **acceleration** (Section 2.3).

List of Transparency Acetates

Figure	Use (✓)	Notes
2.7		
2.9		
2.22		
2.23		

Lecture Demonstrations

Carpenter and Minnix: M-094, M-098, M104, and M-108
Freier and Anderson: Mb-10, 13, 15, 18, 21, and 22
Hilton: M-2, 3, 4, and 5
Meiners: 7-1.2

Laboratory Exercises

#1   Motion in One Dimension
#2   Position, Velocity, and Acceleration

Section 2.1 Displacement

T_{est} = _____ m
T_{act} = _____ m

Section 2.2 Speed and Velocity

T_{est} = _____ m
T_{act} = _____ m

Section 2.3 Acceleration

T_{est} = _____ m
T_{act} = _____ m

Section 2.4 Equations of Kinematics for Constant Acceleration

T_{est} = _____ m
T_{act} = _____ m

Section 2.5 Applications of the Equations of Kinematics

T_{est} = _____ m
T_{act} = _____ m

Section 2.6 Freely Falling Bodies

T_{est} = _____ m
T_{act} = _____ m

Section 2.7 Graphical Analysis of Velocity and Acceleration

T_{est} = _____ m
T_{act} = _____ m

Section 2.8 Concepts & Calculations

T_{est} = _____ m
T_{act} = _____ m

*A*dditional Notes and Ideas   Chapter 2

3   Kinematics in Two Dimensions

Teaching Objectives

- Show that motion in two dimensions is similar to that already learned for the one dimensional case. The motion is described in terms of time and the x and y components of the four vectors: displacement, acceleration, and initial and final velocity.
- Discuss the independence of the motion in each direction that allows us to use the kinematic equations for constant acceleration and analyze the motion in each direction separately.
- Provide several examples of correct usage of the kinematical equations. Discuss the correct usage of plus and minus signs that convey the direction of each component. Reinforce algebraic manipulation of variables to a final state before the insertion of data from the problem.
- Introduce projectile motion as one type of two-dimensional motion with a constant acceleration.

Chapter 5

Concepts at a Glance

In Section 3.2, motion along the x direction and motion along the y direction are independent of each other. As a result, each can be analyzed separately according to the procedures for one-dimensional kinematics discussed in Chapter 2.

In Section 3.3, the horizontal, or x, component of the acceleration is zero in **projectile motion**; and the vertical, or y, component of the acceleration is the **acceleration due to gravity**.

List of Transparency Acetates

Figure	Use (✓)	Notes
3.1		
3.2		
3.5		
3.10		
3.16		

Lecture Demonstrations

Carpenter and Minnix: M-158, M-162, and M-182
Freier and Anderson: Mb-14, 16, 17, 19, 20, 23, 24, and 28
Hilton: M-13
Meiners: 7-2.7 and 7-2.11

Laboratory Exercises

#3A   Projectile Motion Part 1 – Change Initial Speed
#3B   Projectile Motion Part 2 – Change Launch Angle

Section 3.1 Displacement, Velocity, and Acceleration

$T_{est} = $ ____ m
$T_{act} = $ ____ m

Section 3.2 Equations of Kinematics in Two Dimensions

$T_{est} = \underline{}$ m
$T_{act} = \underline{}$ m

Section 3.3 Projectile Motion

$T_{est} = \underline{}$ m
$T_{act} = \underline{}$ m

Section *3.4 Relative Velocity

$T_{est} = \underline{}$ m
$T_{act} = \underline{}$ m

*A*dditional Notes and Ideas   Chapter 3

4 Forces and Newton's Laws of Motion

Teaching Objectives

- Discuss Newton's first law of motion. Define mass and its SI unit, the kilogram. Introduce the concept of an inertial reference frame.
- Present Newton's second law of motion. Introduce force and its SI unit, the newton. Explain the use and construction of free-body diagrams and emphasize their importance in problem solving. Make sure that students understand that the quantity $m\mathbf{a}$ is not a force, but it is equal to the net force acting on the object with mass m. Discuss the vector nature of the second law.
- State Newton's third law of motion. Explain that the forces act on different bodies and that the accelerations that result will usually be different. After providing students with examples of action-reaction pairs, ask them to identify the action-reaction pair in additional examples.
- Introduce the three fundamental forces, then bring the focus to the gravitational force. State Newton's law of universal gravitation. Discuss the weight of an object on earth as the gravitational force that the earth exerts on an object. Make sure students understand the difference between *weight* and *mass*.
- Define surface (normal and frictional) and tension forces and explain that these are not fundamental forces.
- Define equilibrium and provide several examples, using free-body diagrams, applying Newton's laws of motion. Help students understand that the *x, y* axes may be oriented in any convenient direction.

Concepts at a Glance

In Section 4.6, when external forces such as **gravitational** (Section 4.7), **normal** (Section 4.8), **frictional** (Section 4.9), and **tension** (Section 4.10) **forces** act on an object, they must be included in the net force in any application of Newton's second law.

In Section 4.11, both **equilibrium** and **non-equilibrium** problems can be solved with the aid of Newton's second law. For equilibrium problems, the acceleration is zero m/s², while for non-equilibrium problems it is not equal to zero m/s².

List of Transparency Acetates

Figure	Use (✓)	Notes
4.15		
4.17		
4.21		
4.29		
4.30		
4.34		
4.35		

Lecture Demonstrations

Carpenter and Minnix: M-222, 230, 234, 268, 272, 278, and 288
Freier and Anderson: Mc-1 through 5, Md-1 through 4, Mf-2, Mj-2, and Mk-4
Hilton: M-3a and d, M-6 through 8
Meiners: 7-1.6, 8-1.4, 8-1.8, 8-2.2 through 2.4, 8-4.3, and 8-4.7

Laboratory Exercises

#4A   Newton's Second Law Part 1 – Constant Mass
#4B   Newton's Second Law Part 2 – Constant Net Force

Section 4.1 The Concepts of Force and Mass

T_{est} = _____ m
T_{act} = _____ m

Section 4.2 Newton's First Law of Motion

T_{est} = _____ m
T_{act} = _____ m

Section 4.3 Newton's Second Law of Motion

T_{est} = _____ m
T_{act} = _____ m

Section 4.4 The Vector Nature of Newton's Second Law of Motion

T_{est} = _____ m
T_{act} = _____ m

Section 4.5 Newton's Third Law of Motion

T_{est} = _____ m
T_{act} = _____ m

Section 4.6 Types of Forces: an Overview

T_{est} = _____ m
T_{act} = _____ m

Section 4.7 The Gravitational Force

T_{est} = _____ m
T_{act} = _____ m

Section 4.8 The Normal Force

T_{est} = _____ m
T_{act} = _____ m

Section 4.9 Static and Kinetic Frictional Forces

T_{est} = _____ m
T_{act} = _____ m

Section 4.10 The Tension Force

T_{est} = _____ m
T_{act} = _____ m

Section 4.11 Equilibrium Applications of Newton's Laws of Motion

T_{est} = _____ m
T_{act} = _____ m

Section 4.12 Non-Equilibrium Applications of Newton's Laws of Motion

T_{est} = _____ m
T_{act} = _____ m

Section 4.13 Concepts & Calculations

T_{est} = _____ m
T_{act} = _____ m

Additional Notes and Ideas — Chapter 4

5 Dynamics of Uniform Circular Motion

Teaching Objectives
- Define uniform circular motion. Emphasize that the magnitude of the velocity vector is constant, but its direction is not. This naturally leads to the introduction of centripetal acceleration.

Lecture Planning and Notes

- Discuss the radially inward net force that must be present for uniform circular motion. Demonstrate examples of uniform circular motion, describing all of the forces acting on the object. Then, ask students to identify the forces in further examples.
- Use Newton's second law of motion to introduce centripetal force. Point out that in solving problems, the radial component of the net force is found and set equal to mv^2/r.
- Discuss interesting applications involving uniform circular motion such as banked curves, satellite orbits, apparent weightlessness, and vertical circular motion.

Concepts at a Glance

In Section 5.3, uniform circular motion entails a **centripetal acceleration** (Section 5.2) produced by a **net force**, called the **centripetal force**, given by **Newton's second law of motion** (Section 4.3).

List of Transparency Acetates

Figure	Use (✓)	Notes
5.10		
5.11		
5.18		
5.21		

Lecture Demonstrations

Carpenter and Minnix: M-198, 354, and 362
Freier and Anderson: Mf-1 and 2, Mj-2, and Mk-4
Hilton: M-3a and 3d, M-7c, and M-16b and 16f
Meiners: 7-1.6, 8-1.4, and 8-2.5

Laboratory Exercises

#5A Newton's Third Law Part 1 – Collisions
#5B Newton's Third Law Part 2 – Tug-of-War

Section 5.1 Uniform Circular Motion

T_{est} = _____ m
T_{act} = _____ m

Section 5.2 Centripetal Acceleration

T_{est} = _____ m
T_{act} = _____ m

Section 5.3 Centripetal Force

T_{est} = _____ m
T_{act} = _____ m

Section 5.4 Banked Curves

T_{est} = _____ m
T_{act} = _____ m

Section 5.5 Satellites in Circular Orbits

T_{est} = _____ m
T_{act} = _____ m

Section 5.6 Apparent Weightlessness and Artificial Gravity

T_{est} = _____ m
T_{act} = _____ m

Section *5.7 Vertical Circular Motion

T_{est} = _____ m
T_{act} = _____ m

Lecture Planning and Notes   97

Section 5.8 Concepts & Calculations

T_{est} = ____ m
T_{act} = ____ m

*A**dditional Notes and Ideas*   **Chapter 5**

Work and Energy

Teaching Objectives

- Define work. Explain that work is done *on* an object *by* the constant force. Make sure students understand how to select the correct angle between the force and displacement vectors and how this can lead to work being either a positive or a negative quantity. Provide several examples including situations in which several forces are acting on the object, some of which may be perpendicular to the displacement.

- Introduce the work-energy theorem, emphasizing that the net work is used in applying the theorem. Define kinetic energy and its SI unit, the joule. Note for the students that kinetic energy is a scalar quantity and, thus, does not depend on the direction of the velocity of the object. Point out that if negative(positive) work is done on an object, the kinetic energy decreases(increases).
- Discuss the case of work done by the gravitational force and define the gravitational potential energy. Make sure students understand that only the difference in vertical distance is used in the calculation. Explain that the location where the gravitational potential energy is zero may be arbitrarily chosen.
- Define and discuss conservative and nonconservative forces. Make sure that students understand the complete form of the work-energy theorem given as equation 6.7b before proceeding to the case of zero net work done by external nonconservative forces (conservation of mechanical energy).
- Introduce the principle of conservation of mechanical energy. Provide examples and ask students to identify conservative and nonconservative forces acting on the object.
- Introduce average power as the rate at which work is done and define the watt.
- Discuss other forms of energy and state the principle of the conservation of energy.
- If time permits, discuss the optional section on work done by a variable force.

Concepts at a Glance

In Section 6.2, the familiar concept of **work** (Section 6.1) and the new concept of **kinetic energy** (Section 6.2) join with **Newton's second law of motion** (Section 4.3) to produce the **work-energy theorem**.

In Section 6.5, the **total mechanical energy** is formed by combining the concepts of **kinetic energy** (Section 6.2) and **gravitational potential energy** (Section 6.3).

In Section 6.5, an extension is made from the previous figure to illustrate that the **work-energy theorem** leads to the **principle of conservation of mechanical energy** when the net work done by external *nonconservative* forces is zero.

List of Transparency Acetates

Figure	*Use (✓)*	*Notes*
6.2		
6.6		
6.9		
6.17		
6.23		

Lecture Demonstrations
  Carpenter and Minnix: M-406, 414, and 418
  Freier and Anderson: Mn-1 through 3, Mn-6, Mv-1 and 2
  Hilton: M-14a, 14b, and 14e
  Meiners: 9-1.3

Laboratory Exercises
  #6 Work and Energy
  #7 Conservation of Energy

Section 6.1 Work Done by a Constant Force

T_{est} = ____ m
T_{act} = ____ m

Section 6.2 The Work-Energy Theorem and Kinetic Energy

T_{est} = ____ m
T_{act} = ____ m

Section 6.3 Gravitational Potential Energy

T_{est} = ____ m
T_{act} = ____ m

Section 6.4 Conservative Forces versus Nonconservative Forces

T_{est} = ____ m
T_{act} = ____ m

Section 6.5 The Conservation of Mechanical Energy

$T_{est} = \underline{}$ m
$T_{act} = \underline{}$ m

Section 6.6 Nonconservative Forces and the Work-Energy Theorem

$T_{est} = \underline{}$ m
$T_{act} = \underline{}$ m

Section 6.7 Power

$T_{est} = \underline{}$ m
$T_{act} = \underline{}$ m

Section 6.8 Other Forms of Energy and the Conservation of Energy

$T_{est} = \underline{}$ m
$T_{act} = \underline{}$ m

Section 6.9 Work Done by a Variable Force

$T_{est} = \underline{}$ m
$T_{act} = \underline{}$ m

Section 6.10 Concepts & Calculations

$T_{est} = \underline{}$ m
$T_{act} = \underline{}$ m

*A*dditional Notes and Ideas   Chapter 6

7 Impulse and Momentum

Teaching Objectives
- Define impulse, vector quantity equal to the product of the average force and the time during which the force acts. The direction of the impulse is the same as that of the average force. The SI unit for impulse is N•s.
- Define linear momentum. Note that momentum is also a vector quantity that has the same direction as the velocity of the object. Indicate that the total momentum for a system of objects is the vector sum of the individual momenta of the objects.
- Introduce the impulse-momentum theorem by showing how Newton's second law can be rewritten to show that the impulse on an object is equal its change in momentum. Note that this change in momentum is not only dependent on the magnitude of the average force, but also the duration of the application of that force.

- Use Newton's third law of motion to show that two colliding objects exert equal and opposite impulses on each other. If no external forces are present during the collision, this leads to the conservation of linear momentum. Discuss and provide examples of inelastic and elastic collisions. State that in either type of collision that linear momentum is conserved.
- Define the center of mass for two or more objects. Show that the velocity of the center of mass is the same before and after a collision if the total linear momentum of the system remains constant.

Concepts at a Glance

In Section 7.1, two new concepts, **impulse** and **linear momentum**, are introduced and integrated with **Newton's second law of motion** (Section 4.3) to produce the **impulse-momentum theorem** which is used to describe collisions.

In Section 7.2, the **impulse-momentum theorem** (Section 7.1) and **Newton's third law of motion** (Section 4.5) are integrated to obtain the **conservation of linear momentum**.

List of Transparency Acetates

Figure	Use (✓)	Notes
7.1		
7.7		
7.11		
7.13		
7.15		

Lecture Demonstrations

Carpenter and Minnix: M-550, 562, and 566
Freier and Anderson: Mg-4 and 5, Mh-1, and Mi-2
Hilton: M-15
Meiners: 9-4.19

Laboratory Exercises

#8. Impulse versus Change in Momentum
#9A Conservation of Linear Momentum Part 1 – Elastic Collision
#9B Conservation of Linear Momentum Part 2 – Inelastic Collision

Section 7.1 The Impulse-Momentum Theorem

T_{est} = ____ m
T_{act} = ____ m

Section 7.2 The Principle of Conservation of Linear Momentum

T_{est} = ____ m
T_{act} = ____ m

Section 7.3 Collisions in One Dimension

T_{est} = ____ m
T_{act} = ____ m

Section 7.4 Collisions in Two Dimensions

T_{est} = ____ m
T_{act} = ____ m

Section 7.5 Center of Mass

T_{est} = ____ m
T_{act} = ____ m

Section 7.6 Concepts & Calculations

T_{est} = ____ m
T_{act} = ____ m

Additional Notes and Ideas — Chapter 7

8 Rotational Kinematics

Teaching Objectives
- Define angular displacement θ, noting that by convention it is positive for a counterclockwise angular displacement and negative for a clockwise displacement. Indicate that θ increases as a body rotates, even if it is greater than one revolution. Discuss the various units used for angular measurement (degree, rev, and rad) with particular attention to the less familiar and very important SI unit, the radian.

- Define the instantaneous and average rotational velocities. Define instantaneous and average rotational accelerations. Illustrate correct usage of the signs on ω and α through examples of rotating objects in which the signs are the same and in which the signs are opposite.
- Introduce the equations of rotational kinematics for constant angular acceleration. Note the analogy between these equations and those for one-dimensional linear motion. Point out that the equations of rotational kinematics can be used with any self-consistent set of angular units. Also, each equation contains four of the five kinematic variables θ, ω, ω_0, α, and t. Therefore, three variables must be known to find the unknown variable.
- Introduce the relationship between tangential and angular variables for a rigid body rotating about a fixed axis. Make sure students understand that the angular variables must be expressed in radian measure.
- Discuss rolling motion in terms of the angular and tangential variables.
- As an option, introduce the vector nature of the angular velocity and angular acceleration. Point out that the angular velocity vector lies along the rotation axis with the direction determined by a right-hand rule. The direction of the angular velocity vector is determined by the change in the angular velocity.

Concepts at a Glance

In Section 8.2, **time** (Section 1.2) and **angular displacement** (Section 8.1) are combined to produce the concept of **angular velocity**.

In Section 8.2, the concept of **angular acceleration** is the result of bringing together the change in **angular velocity** (Section 8.2) and the **time** (Section 1.2) required for the change to occur.

In Section 8.3, the **equations of rotational kinematics for constant angular acceleration** are obtained by integrating the concepts of **angular displacement, angular velocity, angular acceleration,** and **time**.

List of Transparency Acetates

Figure	Use (✓)	Notes
8.2		
8.12		
8.15		
8.17		
8.18		

Lecture Demonstrations
 Freier and Anderson: Mb-4, Mb-10, Mb-30, and Mr-4
 Hilton: M-16a
 Meiners: 12-2.1

Laboratory Exercises
 #10 Rotational Motion

Section 8.1 Rotational Motion and Angular Displacement

$T_{est} = \underline{}$ m
$T_{act} = \underline{}$ m

Section 8.2 Angular Velocity and Angular Acceleration

$T_{est} = \underline{}$ m
$T_{act} = \underline{}$ m

Section 8.3 The Equations of Rotational Kinematics

$T_{est} = \underline{}$ m
$T_{act} = \underline{}$ m

Section 8.4 Angular Variables and Tangential Variables

$T_{est} = \underline{}$ m
$T_{act} = \underline{}$ m

Section 8.5 Centripetal Acceleration and Tangential Acceleration

$T_{est} = \underline{}$ m
$T_{act} = \underline{}$ m

Section 8.6 Rolling Motion

$T_{est} = \underline{}$ m
$T_{act} = \underline{}$ m

Section *8.7 The Vector Nature of Angular Variables

$T_{est} = \underline{}$ m
$T_{act} = \underline{}$ m

Section 8.8 Concepts & Calculations

$T_{est} = \underline{}$ m
$T_{act} = \underline{}$ m

Additional Notes and Ideas — Chapter 8

Additional Notes and Ideas (*continued*)

9 Rotational Dynamics

Teaching Objectives

- Define torque. Make sure students understand how to select the level arm. Explain that the torque is positive if the force tends to produce a counter-clockwise rotation about the axis and negative if the force tends to produce a clockwise rotation. Note that a given force can produce a larger torque if it located further from the rotation axis.
- Discuss the conditions for a rigid object to be in equilibrium. Point out that the conditions involve the *net force* and *net torque*. Provide examples that illustrate the reasoning strategy for applying the conditions of equilibrium to a rigid object.
- Introduce the concept of center of gravity for a rigid object. Provide examples in which the center of gravity and the torque due to the weight of the object is determined. Discuss the relationship between center of mass and center of gravity.
- Introduce Newton's second law for rotational motion and define rotational inertia. Point out that the law works for an object composed of a number of particles as well as for a single particle in a circular orbit. Discuss the analogy between the rotational and linear forms of Newton's second law. Illustrate the importance of drawing an accurate force-torque diagram for the object to indicate the point(s) at which the force(s) act(are acting) instead of just a simple dot as in the case of most free-body diagrams and to label the axis of rotation.
- Define rotational work and kinetic energy. Discuss the inclusion of the rotational kinetic energy in the total mechanical energy of a rigid object.
- Define angular momentum and state the principle of conservation of angular momentum. Emphasize that conservation is dependent on a zero net external torque acting on the system. Provide examples in which the rotational inertia changes while an object is rotating.

Concepts at a Glance

In Section 9.2, a rigid object is in **equilibrium** when the **sum of the external forces** (Section 4.11) is zero and the **sum the external torques** (Section 9.2) is zero.

In Section 9.4, an object is in **equilibrium** when its **translational acceleration** (Section 4.3) and its **angular acceleration** are zero (Section 9.2). If these accelerations are not zero, then the object is not in equilibrium.

In Section 9.5, the **conservation of mechanical energy** can be applied to an object that has both translational and rotational motion, provided the **rotational kinetic energy** is included in the total mechanical energy along with the **translational kinetic energy** (Section 6.2) and the **gravitational potential energy** (Section 6.3).

In Section 9.6, the **angular momentum** of a system is **conserved** when the net external torque acting on the system is zero N•m. This conservation law is analogous to the **conservation of linear momentum** (Section 7.2) that requires a net external force equal to zero newtons.

List of Transparency Acetates

Figure	Use (✓)	Notes
9.3		
9.7		
9.11		
9.15		
Table 9.1		

Lecture Demonstrations

Carpenter and Minnix: M-614, 622, 646, 650, 662, 670, 682, 768, and 772
Freier and Anderson: Fs-7, Mo-5, Mo-3, Mr-1, Mr-5, Ms-1 through 7, Mt-5, Mt-4 through 7, and Mu-1 through 18
Hilton: M-8b, 10d, 19a, 19b, 19c, 19f through 19i, and 19k
Meiners: 12-3.1, 3.3, 4.3, 4.9, and 5.2

Laboratory Exercises

No laboratory exercises are available for this material.

Section 9.1 The Action of Forces and Torques on Rigid Objects

T_{est} = _____ m
T_{act} = _____ m

Section 9.2 Rigid Objects in Equilibrium

T_{est} = _____ m
T_{act} = _____ m

Section 9.3 Center of Gravity

T_{est} = _____ m
T_{act} = _____ m

Section 9.4 Newton's Second Law for Rotational Motion about a Fixed Axis

T_{est} = _____ m
T_{act} = _____ m

Section 9.5 Rotational Work and Energy

T_{est} = _____ m
T_{act} = _____ m

Section 9.6 Angular Momentum

T_{est} = _____ m
T_{act} = _____ m

Section 9.7 Concepts & Calculations

T_{est} = _____ m
T_{act} = _____ m

*A*dditional Notes and Ideas   Chapter 9

10   Simple Harmonic Motion and Elasticity

Teaching Objectives
- Discuss elastic deformation of objects. Point out that objects have been treated as rigid up until this point, but all objects distort under the application of external forces. Provide examples of linear, shear, and volume deformations.

- Define stress and strain. Explain that if the stress is sufficiently small, the stress is proportional to the strain and the proportionality constant, the modulus, is an intrinsic material property.
- Introduce Hooke's law and the ideal spring. Explain that the spring exerts a restoring force $F = -kx$ on an object attached to the spring when it is displaced from its unstrained length and note the importance of the minus sign.
- Discuss simple harmonic motion and point out that it is described like any motion, in terms of displacement, velocity, and acceleration. Introduce the concepts of amplitude, period, and frequency. Point out the difference between vibrational frequency f and angular frequency ω and write down the relationship between them.
- Define elastic potential energy and discuss the analogy between gravitational and elastic potential energies. Write down the total energy and discuss its conservation. Note that whenever a conservative force acts on a system that a potential energy term is included in its total energy.
- Discuss damped and driven harmonic oscillators. Provide examples including those illustrating resonance. Mention that oscillators lose energy to dissipative forces and note the decay of the amplitude.

Concepts at a Glance

In Section 10.1, the **restoring force of a spring** may contribute to the **net force** that acts on an object. According to **Newton's second law of motion**, the resulting acceleration is directly proportional to the net force.

In Section 10.3, the **elastic potential energy** is added to other energies that may include **translational kinetic energy** (Section 6.2), **rotational kinetic energy** (Section 9.5), and a **gravitational potential energy** (Section 6.3) to give the **total mechanical energy**. The total mechanical energy is conserved if the net work done by external, nonconservative forces is zero (Section 6.5)

List of Transparency Acetates

Figure	Use (✓)	Notes
10.5		
10.6		
10.8		
10.9		
10.11		
10.14		
10.22		
10.25		
10.29		

Lecture Demonstrations
Carpenter and Minnix: M-876 and 892
Freier and Anderson: Ma-8, 9, 10, 12, 13, Mw-3, Mx-1 through 4,
  Mx-6 through 12, My-1, 2, 3, 8, Mz-1, 2, 3, 6, 7, and 9
Hilton: M-14d, 14e, 14f, and 19j
Meiners: 15-1.1, 1.2, 1.3, 1.8, and 1.9

Laboratory Exercises
#11 Hooke's Law
#12 Simple Harmonic Motion, Mass on a String
#13 Simple Harmonic Motion, Simple Pendulum

Section 10.1 The Ideal Spring and Simple Harmonic Motion

T_{est} = ____ m
T_{act} = ____ m

Section 10.2 Simple Harmonic Motion and the Reference Circle

T_{est} = ____ m
T_{act} = ____ m

Section 10.3 Energy and Simple Harmonic Motion

T_{est} = ____ m
T_{act} = ____ m

Section 10.4 The Pendulum

T_{est} = ____ m
T_{act} = ____ m

114 INSTRUCTOR'S RESOURCE GUIDE — Chapter 5

Section 10.5 Damped Harmonic Motion

T_{est} = ____ m
T_{act} = ____ m

Section 10.6 Driven Harmonic Motion and Resonance

T_{est} = ____ m
T_{act} = ____ m

Section 10.7 Elastic Deformation

T_{est} = ____ m
T_{act} = ____ m

Section 10.8 Stress, Strain, and Hooke's Law

T_{est} = ____ m
T_{act} = ____ m

Section 10.9 Concepts & Calculations

T_{est} = ____ m
T_{act} = ____ m

Additional Notes and Ideas — Chapter 10

Additional Notes and Ideas (*continued*)

11 Fluids

Teaching Objectives

- Introduce fluids as materials that can flow including gases as well as liquids.
- Define mass density and specific gravity.
- Define pressure and its SI unit, the pascal (Pa). Discuss the measurement of pressure and other pressure units. Make sure students understand the difference between gauge and absolute pressure.
- Discuss the relationship between pressure and depth within a fluid. Note that the pressure depends on the vertical distance below the surface and that all points at a given depth are at the same pressure.
- Present Pascal's principle and provide several examples of its application. Point out that the relation $F_1/A_1 = F_2/A_2$ applies only when points 1 and 2 are at the same depth in the fluid.
- Introduce the buoyant force and Archimedes' principle. Note that the density of the displaced fluid is used, not the density of the object.
- Discuss fluids in motion differentiating between the following: steady and unsteady flow, compressible and incompressible fluids, viscous and nonviscous flow, and rotational and irrotational flow. Illustrate the use of streamlines.
- Define mass flow rate and write down the equation of continuity as an expression of the conservation of mass.
- Introduce Bernoulli's equation in terms of the work-energy theorem. Provide several examples of its usage.
- (optional) Discuss viscosity and introduce Poiseuille's law for viscous fluid flow.

Concepts at a Glance

In Section 11.3, the **gravitational force** (Section 4.7) and the **collisional forces exerted by a fluid** (Section 11.2) are used with **Newton's second law of motion** to determine the relation between pressure and depth in a static fluid. This figure is expanded from Figures 4.26 and 10.12.

In Section 11.9, the **work-energy theorem** (Sections 6.2 and 6.5) leads to **Bernoulli's equation** when the net work done by external forces is not zero.

List of Transparency Acetates

Figure	Use (✓)	Notes
11.6		
11.15		
11.18		
11.30		
11.31		
11.33		
11.34		
11.35		
11.37		

Lecture Demonstrations

Carpenter and Minnix: F-005, 015, 025, 035, 115, 120, 130, 215, 225, and 235
Freier and Anderson: Fa through Ff, Fj-1 through Fj-11, Fk-2, and Fl-1
Hilton: M-20b, 20c, and 22a through 22d
Meiners: 16-2.2, 2.5, 2.6, 4.5, 4.6, 4.11, 17-2.5, and 17-2.12

Laboratory Exercises

#14   Buoyant Force

Section 11.1 Mass Density

$T_{est} = \underline{}$ m
$T_{act} = \underline{}$ m

Lecture Planning and Notes 117

Section 11.2 Pressure

T_{est} = ____ m
T_{act} = ____ m

Section 11.3 Pressure and Depth in a Static Fluid

T_{est} = ____ m
T_{act} = ____ m

Section 11.4 Pressure Gauges

T_{est} = ____ m
T_{act} = ____ m

Section 11.5 Pascal's Principle

T_{est} = ____ m
T_{act} = ____ m

Section 11.6 Archimedes' Principle

T_{est} = ____ m
T_{act} = ____ m

Section 11.7 Fluids in Motion

T_{est} = ____ m
T_{act} = ____ m

Section 11.8 The Equation of Continuity

T_{est} = ____ m
T_{act} = ____ m

Section 11.9 Bernoulli's Equation

T_{est} = _____ m
T_{act} = _____ m

Section 11.10 Applications of Bernoulli's Equation

T_{est} = _____ m
T_{act} = _____ m

Section *11.11 Viscous Flow

T_{est} = _____ m
T_{act} = _____ m

Section 11.12 Concepts & Calculations

T_{est} = _____ m
T_{act} = _____ m

*A*dditional Notes and Ideas — Chapter 11

Additional Notes and Ideas (*continued*)

12 Temperature and Heat

Teaching Objectives

- Introduce common temperature scales and the measurement of temperature. Demonstrate the conversion of temperatures between different scales.
- Discuss thermal expansion of objects and provide examples including one that illustrates the effect of thermal stress. Demonstrate area expansion, showing that holes also expand with increasing temperature.
- Define heat as a transfer of energy from a higher temperature region or object to a lower temperature region or object due to the difference in temperatures. Define the internal energy of a substance. Distinguish between heat and internal energy, noting that heat is not a new type of energy. Discuss the common energy units: joule (J), calorie (cal), Calorie (kcal), and British thermal unit (btu).
- Define the specific heat capacity. Discuss the difference the in specific heat capacity of gases when measured at constant volume and at constant pressure. Note that the difference is due to the positive work done by the system when the temperature is increased at constant pressure. Introduce calorimetry.
- Discuss phase changes that occur in materials that result from the addition or removal of heat. Define latent heats of vaporization, fusion, and sublimation. Provide calorimetric examples involving phase changes.
- (optional) Discuss equilibrium between material phases. Introduce the vapor pressure curve. Discuss relative humidity.

Concepts at a Glance

In Section 12.6, when **heat** is added or removed from an object, its **internal energy** can change. This change can cause a **change in temperature** (Section 12.7) or a **change in phase** (Section 12.8).

120 INSTRUCTOR'S RESOURCE GUIDE — Chapter 5

List of Transparency Acetates

Figure	Use (✓)	Notes
12.1		
12.10		
12.15		
12.17		
12.25		
12.27		
12.31		
12.32		

Lecture Demonstrations
  Carpenter and Minnix: H-010, 014, 018, 040, 064, 068, 220, 230, 240, and 264
  Freier and Anderson: Ha-1 through Ha-12, Hb-1, Hb-2, Hj-1, 4, 7, 8,
    Hk-1, 3, 7, 9, 10, 11, and Hm-4
  Hilton: H-1, 2, 5d and 5e
  Meiners: 25-2.1 through 2.3, 27-3.1, and 27-3.6

Laboratory Exercises
  #15   Temperature and Heat
  #16   Specific Heat

Section 12.1 Common Temperature Scales

$T_{est} = \_\_\_\_$ m
$T_{act} = \_\_\_\_$ m

Section 12.2 The Kelvin Temperature Scale

$T_{est} = \_\_\_\_$ m
$T_{act} = \_\_\_\_$ m

Section 12.3 Thermometers

T_{est} = ____ m
T_{act} = ____ m

Section 12.4 Linear Thermal Expansion

T_{est} = ____ m
T_{act} = ____ m

Section 12.5 Volume Thermal Expansion

T_{est} = ____ m
T_{act} = ____ m

Section 12.6 Heat and Internal Energy

T_{est} = ____ m
T_{act} = ____ m

Section 12.7 Heat and Temperature Change: Specific Heat Capacity

T_{est} = ____ m
T_{act} = ____ m

Section 12.8 Heat and Phase Change: Latent Heat

T_{est} = ____ m
T_{act} = ____ m

*Section *12.9 Equilibrium between Phases of Matter*

$T_{est} = \_\_\_\_ m$
$T_{act} = \_\_\_\_ m$

*Section *12.10 Humidity*

$T_{est} = \_\_\_\_ m$
$T_{act} = \_\_\_\_ m$

Section 12.11 Concepts & Calculations

$T_{est} = \_\_\_\_ m$
$T_{act} = \_\_\_\_ m$

Additional Notes and Ideas — Chapter 12

Additional Notes and Ideas (*continued*)

13 The Transfer of Heat

Teaching Objectives

- Introduce convection as a means of heat transfer. Distinguish between natural and forced convection.
- Discuss thermal conduction and note that bulk motion of material does not play a role in the heat transfer. Define the thermal conductivity of the material. Discuss and provide examples of thermal conductors and thermal insulators. Point out that *if the ends of a bar are maintained* at different temperatures, a *steady* transfer of heat occurs from the warmer end to the cooler end.
- Discuss radiation as another process for energy transfer. Point out that all objects continuously absorb and emit electromagnetic radiation. Note that good absorbers are good emitters and poor absorbers are poor emitters. Define the perfect blackbody. Present the Stefan-Boltzmann law and point out that the absolute temperature must be used.

Concepts at a Glance

In Section 13.1, **heat** (Section 12.6) is transferred from one place to another by three methods: **convection** (Section 13.1), **conduction** (Section 13.2), and **radiation** (Section 13.3). As discussed in Chapter 12, heat can change the **internal energy** (Section 12.6) of a substance, resulting in a **change in the temperature** (Section 12.7) or a **change in the phase** (Section 12.8) of the substance.

List of Transparency Acetates

Figure	Use (✓)	Notes
13.4		
13.8		
13.9		
13.15		

Lecture Demonstrations

Carpenter and Minnix: H-160, 140, and 144
Freier and Anderson: Hc-1, Hc-2, Hd-1 through 7, and Hf-1 through 5
Hilton: H-3a through 3c
Meiners: 26-3.1, 3.2, 3.4, 3.6, 3.8, 38-5.1, 5.3, and 5.4

Laboratory Exercises

No laboratory exercises are available for this material.

Section 13.1 Convection

$T_{est} = \_\_\_\_$ m
$T_{act} = \_\_\_\_$ m

Section 13.2 Conduction

$T_{est} = \_\_\_\_$ m
$T_{act} = \_\_\_\_$ m

Section 13.3 Radiation

$T_{est} = \_\_\_\_$ m
$T_{act} = \_\_\_\_$ m

Section 13.4 Applications

T_{est} = _____ m
T_{act} = _____ m

Section 13.5 Concepts & Calculations

T_{est} = _____ m
T_{act} = _____ m

Additional Notes and Ideas — Chapter 13

14 The Ideal Gas Law and Kinetic Theory

Teaching Objectives

- Define atomic and molecular masses. Introduce the mole and Avogadro's number. Explain relationships between these quantities and provide example calculations using them.
- State the ideal gas law in both forms: $PV = nRT$ and $PV = NkT$. Point out that the ideal gas theory applies to real gases when their densities are sufficiently small. Note that absolute pressure and Kelvin temperature are used in the formula.
- Present the kinetic theory of gases as an application of Newton's second and third laws of motion and the ideal gas law. Discuss the distribution of molecular speeds. Distinguish between average and *rms* speeds. Write down the relationship between average translational kinetic energy of particles in the gas and the temperature. Note that the average kinetic energy is not dependent on molecular mass, pressure, or volume. Introduce the internal energy of a monatomic ideal gas.
- (optional) Discuss gaseous diffusion and write down Fick's law as an analogy to thermal diffusion.

Concepts at a Glance

In Section 14.3, to formulate the **kinetic theory of gases, Newton's second** (Section 4.3) and **third** (Section 4.5) **laws of motion** are brought together with the **ideal gas law** (Section 14.2).

List of Transparency Acetates

Figure	Use (✓)	Notes
14.6		
14.9		
14.14		
14.16		

Lecture Demonstrations

Carpenter and Minnix: H-440 and 450
Freier and Anderson: Hg-1, 2, 4, and Hh-1 through 5
Hilton: H-5f
Meiners: 27-2.1, 2.7, 2.8, 7.1, 7.5, and 7.6

Laboratory Exercises
   #14   Ideal Gas Law

Section 14.1 Molecular Mass, the Mole, Avogadro's Number

T_{est} = ____ m
T_{act} = ____ m

Section 14.2 The Ideal Gas Law

T_{est} = ____ m
T_{act} = ____ m

Section 14.3 Kinetic Theory of Gases

T_{est} = ____ m
T_{act} = ____ m

Section *14.4 Diffusion

T_{est} = ____ m
T_{act} = ____ m

Section 14.5 Concepts & Calculations

T_{est} = ____ m
T_{act} = ____ m

Additional Notes and Ideas Chapter 14

15 Thermodynamics

Teaching Objectives
- Introduce thermodynamics as the branch of physics dealing with the laws of energy in the form of heat and work. Define the following concepts: system, surroundings, adiabatic and diathermal walls, and state of a system. Note that diathermal walls allow thermal contact without an exchange of particles.
- State the zeroth law of thermodynamics, defining thermal equilibrium. The measurement of temperature is a good example of an application. This is the reason why temperature can be considered a property of an object.

Lecture Planning and Notes 129

- Discuss the first law of thermodynamics, as expression of the conservation of energy. Point out that internal energy is a *function of state* and that pressure and volume are not functions of state. Discuss what happens when different paths are taken between the initial and final states of a system.
- Provide examples that apply the first law of thermodynamics to the following types of processes: isobaric, isochoric, isothermal, adiabatic, and cyclical (reversible and irreversible). Discuss thermal processes involving ideal gases. Make sure that students understand the sign convention used for the heat and work when using the first law of thermodynamics.
- State the second law of thermodynamics regarding the natural flow of heat from a substance at higher temperature to one at lower temperature. Indicate that the reverse, flow from a substance at lower temperature to a substance at higher temperature, does not occur spontaneously. Introduce the following applications: heat engines, Carnot engine, refrigerators, air conditioners, and heat pumps. Define efficiency, Carnot efficiency, and coefficient of performance.

Concepts at a Glance

In Section 15.3, the concepts of **work** (Sections 6.1 and 6.9), **heat** (Section 12.6), and the **internal energy** (Section 12.6) of a system are related by the **first law of thermodynamics**, an expression of the law of conservation of energy.

In Section 15.7, the **first and second laws of thermodynamics** (Sections 15.3 and 15.7, respectively) are used to evaluate the performance of **heat engines, refrigerators, air conditioners, and heat pumps** (Sections 15.8-10).

List of Transparency Acetates

Figure	Use (✓)	Notes
15.2		
15.5		
15.9		
15.10		
15.13		
15.14		
15.15		
15.17		
15.21		

Lecture Demonstrations
Carpenter and Minnix: H-320, 340, 395, 405, and 500
Freier and Anderson: He-1 through 6, Hm-1, 2, and 5
Hilton: H-5a and 5b
Meiners: 26-4.1, 4.5, and 4.6

Laboratory Exercises
No laboratory exercises are available for this material.

Section 15.1 Thermodynamic Systems and Their Surroundings

T_{est} = ____ m
T_{act} = ____ m

Section 15.2 The Zeroth Law of Thermodynamics

T_{est} = ____ m
T_{act} = ____ m

Section 15.3 The First Law of Thermodynamics

T_{est} = ____ m
T_{act} = ____ m

Section 15.4 Thermal Processes

T_{est} = ____ m
T_{act} = ____ m

Section 15.5 Thermal Processes Using an Ideal Gas

T_{est} = ____ m
T_{act} = ____ m

Section 15.6 Specific Heat Capacities

T_{est} = ____ m
T_{act} = ____ m

Section 15.7 The Second Law of Thermodynamics

T_{est} = ____ m
T_{act} = ____ m

Section 15.8 Heat Engines

T_{est} = ____ m
T_{act} = ____ m

Section 15.9 Carnot's Principle and the Carnot Engine

T_{est} = ____ m
T_{act} = ____ m

Section 15.10 Refrigerator's, Air Conditioners, and Heat Pumps

T_{est} = ____ m
T_{act} = ____ m

Section 15.11 Entropy

T_{est} = ____ m
T_{act} = ____ m

Section 15.12 The Third Law of Thermodynamics

T_{est} = ____ m
T_{act} = ____ m

Section 15.13 Concepts & Calculations

T_{est} = ____ m
T_{act} = ____ m

Additional Notes and Ideas — Chapter 15

Lecture Planning and Notes

16 Waves and Sound

Teaching Objectives

- Define a *wave* as a traveling disturbance that carries energy from one place to another. Indicate that individual particles have very limited motion. Demonstrate one-dimensional waves by using a taut string. Show a two-dimensional wave resulting from the drop of a pebble into a pan of water. Discuss how waves are produced and define *periodic waves*.

- Discuss the difference between *traverse* and *longitudinal* waves. Demonstrate both types of waves on a stretched spring.

- Define *wavelength*. Remind students of the concepts of period, cycle, amplitude, and frequency learned in Chapter 10. Show students that sinusoidal wave travels one wavelength in one period.

- Introduce the relationship between the speed of a wave and its wavelength and frequency. State that the formula, $v = \lambda f$, may be used with any kind of wave.

- Define *sound* as a longitudinal wave composed of alternating regions of condensations (greater pressure than normal) and rarefaction (less pressure than normal). Discuss the difference between this definition and the interpretation of sound in the brain.

- Point out that sound requires a medium; and that the speed of a sound wave is determined by the inertia and elasticity of the medium. Provide formulas for estimating the speed of sound in solids, liquids, and ideal gases and discuss the differences between these media. Remark that the temperature in the formula for the speed of sound in ideal gases is the Kelvin temperature. Discuss how the wavelength and speed of a sound wave change as it moves from one medium to another, but the frequency remains the same.

- Discuss how waves carry energy and remind students of the definition of power. Define intensity and intensity level. Point out the difference between power and intensity.

- Demonstrate the Doppler effect and point out that the frequency decreases when the source is moving away from the observer and increases when the source is moving toward the observer. State that similar effects occur when the observer is moving toward or away from the source. Discuss what happens when both the source and the observer are moving relative to one another.

Concepts at a Glance

In Section 16.2, for periodic waves, the terms **cycle**, **amplitude**, **period**, and **frequency** have the same meaning as they do in **simple harmonic motion** (Section 10.4).

In Section 16.7, the concept of **sound intensity** takes into account both the **sound power** and the **area** through which the power passes.

In Section 16.9, the **Doppler effect** arises when the source and the observer of a sound wave have different velocities with respect to the medium through which the sound travels.

List of Transparency Acetates

Figure	Use (✓)	Notes
16.2		
16.3		
16.6		
16.11		
16.12		
16.18		
16.29		
16.30		
16.31		

Lecture Demonstrations

Carpenter and Minnix: W-005, 010, 095, 305, and 380
Freier and Anderson: Sa-3 through 6, 12, 13, 14, Sh-1 through 3, Si-1 through 3, and Sl-1
Hilton: S-2a, 2c, 2d, 3f, 3g, and 6
Meiners: 18-3.1, 19-6.1, and 19-6.2

Laboratory Exercises

There are no available exercises for this material.

Section 16.1 The Nature of Waves

T_{est} = ____ m
T_{act} = ____ m

Section 16.2 Periodic Waves

$T_{est} = \underline{} m$
$T_{act} = \underline{} m$

Section 16.3 The Speed of a Wave on a String

$T_{est} = \underline{} m$
$T_{act} = \underline{} m$

Section *16.4 The Mathematical Description of a Wave

$T_{est} = \underline{} m$
$T_{act} = \underline{} m$

Section 16.5 The Nature of Sound

$T_{est} = \underline{} m$
$T_{act} = \underline{} m$

Section 16.6 The Speed of Sound

$T_{est} = \underline{} m$
$T_{act} = \underline{} m$

Section 16.7 Sound Intensity

$T_{est} = \underline{} m$
$T_{act} = \underline{} m$

Section 16.8 Decibels

$T_{est} = \_\_\_\_$ m
$T_{act} = \_\_\_\_$ m

Section 16.9 The Doppler Effect

$T_{est} = \_\_\_\_$ m
$T_{act} = \_\_\_\_$ m

Section 16.10 Applications of Sound in Medicine

$T_{est} = \_\_\_\_$ m
$T_{act} = \_\_\_\_$ m

Section *16.11 The Sensitivity of the Human Ear

$T_{est} = \_\_\_\_$ m
$T_{act} = \_\_\_\_$ m

Section 16.12 Concepts & Calculations

$T_{est} = \_\_\_\_$ m
$T_{act} = \_\_\_\_$ m

Additional Notes and Ideas — Chapter 16

17 The Principle of Linear Superposition and Interference Phenomena

Teaching Objectives

- Introduce the principle of linear superposition and define constructive and destructive interference. Illustrate the concept of phase with a demonstration. Make sure that students understand that displacements add together, not intensities. Explain how a phase difference can result when two initially identical and in-phase waves travel different distances to meet at the same point.
- Discuss diffraction in terms of the principle of linear superposition; and show that your voice can still be heard even though you are behind some obstacle such as a wall. Point out that the extent of diffraction is larger when the ratio of λ/D is larger than when the ratio is smaller.

- Discuss the phenomenon of beats in terms of linear superposition of two waves with slightly different frequencies. Provide a demonstration of the phenomenon. Point out that musical instruments may be tuned by listening for beats and subsequently minimizing the beat frequency.
- Introduce the concept of standing waves. Illustrate both transverse and longitudinal standing waves. Make sure students can point out nodes and antinodes and understand how to count them. Point out that the distance between two successive nodes (or antinodes) is equal to one-half of a wavelength. Discuss natural frequency and the resonant condition. Provide formulas for determining the natural frequencies of strings and open tubes. Demonstrate how these are used to make musical instruments.

Concepts at a Glance

In Section 17.3, we can understand the interference phenomenon of **diffraction** by combining our knowledge of **sound waves** (Section 16.5) with the **principle of linear superposition** (Sections 17.1 and 17.2).

In Section 17.4, Figure 17.9 is extended as the **principle of linear superposition** (Sections 17.1 and 17.2) is used to explain he phenomenon of **beats**.

In Section 17.5, Figure 17.14 is extended to indicate that **transverse and longitudinal standing waves** are also related to the **principle of linear superposition** (Sections 17.1 and 17.2).

List of Transparency Acetates

Figure	Use (✓)	Notes
17.3		
17.4		
17.10		
17.16		
17.18		
17.23		
17.25		

Lecture Demonstrations

Carpenter and Minnix: W-105, 145, 150, 170, 210, 230, 260, 325, 330, 335
Freier and Anderson: Sa-8, 17, 18, Se-1 through 5, 8 through 11, Si-4 through 6, and Sl-3
Hilton: S-2h, 4b, 4c, and 5
Meiners: 18-5.6, 5.7, 7.1, 19-3.1, 3.3, 3.4, 3.5, 5.4, and 5.5

Lecture Planning and Notes

Laboratory Exercises
- #18 Superposition
- #19 Interference in Sound - Beats

Section 17.1 The Principle of Linear Superposition

$T_{est} = \_\_\_\_$ m
$T_{act} = \_\_\_\_$ m

Section 17.2 Constructive and Destructive Interference of Sound Waves

$T_{est} = \_\_\_\_$ m
$T_{act} = \_\_\_\_$ m

Section 17.3 Diffraction

$T_{est} = \_\_\_\_$ m
$T_{act} = \_\_\_\_$ m

Section 17.4 Beats

$T_{est} = \_\_\_\_$ m
$T_{act} = \_\_\_\_$ m

Section 17.5 Transverse Standing Waves

$T_{est} = \_\_\_\_$ m
$T_{act} = \_\_\_\_$ m

Section 17.6 Longitudinal Standing Waves

$T_{est} = \_\_\_\_$ m
$T_{act} = \_\_\_\_$ m

140 INSTRUCTOR'S RESOURCE GUIDE					Chapter 5

*Section *17.7 Complex Sound Waves*

T_{est} = _____ m
T_{act} = _____ m

Section 17.8 Concepts & Calculations

T_{est} = _____ m
T_{act} = _____ m

# *A*dditional Notes and Ideas		Chapter 17

18 Electric Forces and Electric Fields

Teaching Objectives

- State that there are two kinds of charge; and these have been labeled positive and negative. Explain that the SI unit for charge is the coulomb (C); and mention the value for e. State that charge is known to be quantized and discuss the conservation of charge.
- Discuss the difference between electrical conductors and insulators. Demonstrate that objects can become charged through contact and by induction. Point out that most objects are electrically neutral, but this does not mean an absence of charged particles; and that the charge imbalance is usually small, but obviously significant.
- Remind students of the unit prefixes commonly used, such as $1 \ \mu C = 1 \times 10^{-6}$ C.
- Show that for objects with like charges, there is a force of repulsion between them while for objects with unlike charges, a force of attraction exists. Discuss the experimental evidence that leads to Coulomb's law. Point out that the force is directed along a line connecting the two point charges. Discuss the similarities and differences between Newton's law of gravitation and Coulomb's law. Illustrate using Coulomb's law with an arrangement of three or more point charge.
- Point out the use of the absolute value symbol used for the charges when applying Coulomb's law.
- Introduce the concept of a test charge and define the electric field with its SI unit (N/C). Make sure students understand that a charge creates an electric field at all points in space; and in turn, any other charges present are subjected to the electrostatic force as a result of interacting with the electric field. Discuss the electric field inside a parallel plate capacitor. Illustrate the usefulness of electric field lines in visualizing the strength and behavior of the electric field. Point out that the electric field points radially outward from a positive point charge and radially inward for negative point charges.
- Introduce the concepts of electric flux and the imaginary Gaussian closed surface. State that it can be of any convenient shape. Point out that it is the electric field component in the direction normal to the surface that is used in Gauss' Law.

Concepts at a Glance

In Section 18.2, any **external electrostatic forces** that act on an object must be included when determining the **net external force** that is used in **Newton's second law of motion** (Section 4.3).

In Section 18.6, a **test charge** and the **electrostatic force** (Section 18.5) that acts on it are brought together to define the concept of an **electric field**.

In Section 18.9, Figure 18.16 (Section 18.6) is extended as the **electric field** and the **surface** through which it passes are brought together to define the concept of **electric flux**. **Gauss' law** in formulated in terms of electric flux.

List of Transparency Acetates

Figure	Use (✓)	Notes
18.3		
18.17		
18.22		
18.23		
18.24		
18.26		
18.27		
18.30		
18.34		

Lecture Demonstrations
Carpenter and Minnix: E-040, 060, 065, 085, and 090
Freier and Anderson: Ea-1, 2, 5, 6, 8, 11, 12, 14, 15, 17, Eb-1, 3, 4, 9, 10, 12, and Ec-2 through 6
Hilton: E-1a through 1g and E-5b
Meiners: 29-1.4, 1.9, 1.18, 1.23, and 2.1

Laboratory Exercises
No laboratory exercises are available for this material.

Section 18.1 The Origin of Electricity

T_{est} = ____ m
T_{act} = ____ m

Section 18.2 Charged Objects and the Electric Force

T_{est} = _____ m
T_{act} = _____ m

Section 18.3 Conductors and Insulators

T_{est} = _____ m
T_{act} = _____ m

Section 18.4 Charging by Contact and by Induction

T_{est} = _____ m
T_{act} = _____ m

Section 18.5 Coulomb's Law

T_{est} = _____ m
T_{act} = _____ m

Section 18.6 The Electric Field

T_{est} = _____ m
T_{act} = _____ m

Section 18.7 Electric Field Lines

T_{est} = _____ m
T_{act} = _____ m

Section 18.8 The Electric Field inside a Conductor: Shielding

$T_{est} = \underline{}\,m$
$T_{act} = \underline{}\,m$

Section 18.9 Gauss' Law

$T_{est} = \underline{}\,m$
$T_{act} = \underline{}\,m$

*Section *18.10 Copiers and Computer Printers*

$T_{est} = \underline{}\,m$
$T_{act} = \underline{}\,m$

Section 18.11 Concepts & Calculations

$T_{est} = \underline{}\,m$
$T_{act} = \underline{}\,m$

Additional Notes and Ideas — Chapter 18

Additional Notes and Ideas (*continued*)

19 Electric Potential Energy and the Electric Potential

Teaching Objectives

- Carefully introduce and distinguish the concepts of electric potential and electric potential energy and their respective units. Indicate the importance of the sign on the potential. Note that the electric field points from a region of higher potential to one of lower potential and that a positive test charge is repelled by the higher potential region. Indicate the differences in behavior for positive and negative test charges.
- Indicate the conservative nature of the electrostatic force and note that the work done by it is path independent.
- Point out that the electric potential is a scalar quantity and that its zero may be chosen at any convenient location. Normally, we choose the potential to be zero at an infinite distance away from a charge in calculating the electric potential due to that charge. Note that only potential differences have physical meaning and can be measured. Demonstrate the use of a voltmeter.
- Provide examples of the calculation of the electric potential at a point due to the presence of two or more charges.
- Define the electron volt (eV) (= 1.60×10^{-19} J) as the change in electric potential energy of an electron as it moves through a potential difference of one volt. Make sure students understand that electric potential energy is part of the total energy of an object and that it can be converted to kinetic energy.
- Define the equipotential surface. Illustrate equipotential surfaces for an isolated point charge, between two charges with the same sign, between two charges with opposite sign, and for various charge distributions. Note that the electric force does no work in moving a charge from one point to another on an equipotential surface; and therefore, the force and the electric field must be perpendicular to the surface.

- Introduce capacitors and define the capacitance and its SI unit, the farad (F). Point out that capacitors used in most everyday electronic devices have capacitances in the range from picofarad to microfarad. Discuss the effect of a dielectric material and define the dielectric constant. Discuss in detail the parallel plate capacitor and the use of capacitors as energy storage devices. Provide examples of devices that use capacitors, such as those used in biomedical applications.

Concepts at a Glance

In Section 19.2, the **electric potential energy** (Section 19.1) and the notion of a **test charge** (Section 18.6) are combined to produce the idea of an **electric potential**.

In Section 19.2, the **electric potential energy** is another type of energy that an object can have. The **total energy** is **conserved** when nonconservative forces are absent or do no net work. *This chart is a continuation of Figures 6.16 (Section 6.5), 9.23 (Section 9.5), and 10.27 (Section 10.5).*

List of Transparency Acetates

Figure	Use (✓)	Notes
19.1		
19.2		
19.14		
19.15		
19.18		
19.20		

Lecture Demonstrations

Carpenter and Minnix: E-160, 180, 210, and 240
Freier and Anderson: Ea-7, 18, 22, 23, Eb-7, 8, Ec-1, Ed-1 through 4, 7, and 8
Hilton: E-1h through 1j and E-4b through 4d
Meiners: 29-1.25, 1.26, 2.8, 4.1, and 4.13

Laboratory Exercises

There are no exercises for this material.

Section 19.1 Potential Energy

T_{est} = ____ m
T_{act} = ____ m

Section 19.2 The Electric Potential Difference

T_{est} = ____ m
T_{act} = ____ m

Section 19.3 The Electric Potential Difference Created by Point Charges

T_{est} = ____ m
T_{act} = ____ m

Section 19.4 Equipotential Surfaces and Their Relation to the Electric Field

T_{est} = ____ m
T_{act} = ____ m

Section 19.5 Capacitors and Dielectrics

T_{est} = ____ m
T_{act} = ____ m

Section *19.6 Biomedical Applications of Electric Potential Differences

T_{est} = _____ m
T_{act} = _____ m

Section 19.7 Concepts & Calculations

T_{est} = _____ m
T_{act} = _____ m

Additional Notes and Ideas — Chapter 19

20 Electric Circuits

Teaching Objectives

- Introduce the concept of the electromotive force, emf. Describe batteries and the symbol used to represent them in electric circuits.
- Discuss the effect of connecting a battery to an electric circuit. State that an electric field exists within and parallel to the wire and points from the positive terminal toward the negative terminal. Point out that free electrons are accelerated by the electric field. Draw an imaginary surface perpendicular to the movement of the electrons and define the electric current, I, as the number of electrons passing through the surface, Δq, per unit time, Δt. Introduce the SI unit for current, the ampere (A). Describe direct and alternating currents.
- Introduce Ohm's law, $R = V/I$, to define resistance and its SI unit, the ohm $\Omega = $ V/A. Point out that many, but not all, materials obey Ohm's law. Define resistivity as an intrinsic property of the material and graphically illustrate its temperature dependence.
- Note that since electrons are moving from a point of higher potential to a point of lower potential, energy is being transferred from the battery to any devices connected to it. This energy is converted to kinetic energy if the device is a motor or to internal energy if the device is a resistor. This energy is transferred at a rate given by $P = IV = I^2R = V^2/R$..
- Discuss alternating voltage sources and introduce the symbol used in electric circuits. Discuss alternating current in a circuit containing only resistance and define *rms* voltage and current. Introduce the concept of average power, $\overline{P} = I_{rms}V_{rms}$.
- Introduce the concept of equivalent resistance for resistors in series and in parallel. Make sure students understand the mathematics of taking reciprocals, especially on a calculator. Point out the existence of the internal resistance of a battery and the difference between the terminal voltage and the emf.
- List Kirchoff's rules and provide several examples of their usage. Point out the inclusion of the conservation of charge and the conservation of energy concepts within Kirchoff's rules.
- Demonstrate devices that are used to measure current, voltage, and resistance (optional) and explain how they function. Point out the differences between analog and digital devices.
- Introduce the concept of equivalent capacitance for capacitors in series and in parallel. Discuss *RC* circuits and provide several examples. Make sure students understand the concept of time constant and the charging/discharging behavior of the capacitors in the circuit.

- If time permits, cover optional Section 20.14 on safety and physiological effects of current. This may be one of the (many) practical things that students can learn in their physics class that they will not encounter, in general, elsewhere.

Concepts at a Glance

In Section 20.1, the concept of **electric current** is formulated from the concept of **electric charge** (Section 18.1) and the **time** during which it flows in a circuit.

In Section 20.2, the concepts of **electric potential difference, V** (Section 19.2) and the **electric current** (Section 20.1) are brought together to define the **electrical resistance**. **Ohm's law** specifies that the resistance remains constant as the voltage and current change.

In Section 20.10, **Kirchoff's rules** are applications to electric circuits of the law of **conservation of electric charge** and the principle of **conservation of energy** (Section 6.8). This figure also extends Figures 19.3 and 20.4.

List of Transparency Acetates

Figure	Use (✓)	Notes
20.2		
20.7		
20.10		
20.19		
20.20		
20.23		
20.24		
20.36		
20.37		

Lecture Demonstrations

Carpenter and Minnix: E-300, 360, 450
Freier and Anderson: Ed-6 through 8, Ee-2, 3, 4, Eg-1 through 7, Eh-1, 3, 4, and Eo-1, 2, 3, 5, 6, 7, and 12
Hilton: E-2b, 2c, 3, 3a, 3b, 3c, 3d, and 3g
Meiners: 30-1.4

Laboratory Exercises
- #20 Ohm's Law
- #21 DC Series Wiring
- #22 DC Parallel Wiring

Section 20.1 Electromotive Force and Current

T_{est} = ____ m
T_{act} = ____ m

Section 20.2 Ohm's Law

T_{est} = ____ m
T_{act} = ____ m

Section 20.3 Resistance and Resistivity

T_{est} = ____ m
T_{act} = ____ m

Section 20.4 Electric Power

T_{est} = ____ m
T_{act} = ____ m

Section 20.5 Alternating Current

$T_{est} = \_\_\_\_$ m
$T_{act} = \_\_\_\_$ m

Section 20.6 Series Wiring

$T_{est} = \_\_\_\_$ m
$T_{act} = \_\_\_\_$ m

Section 20.7 Parallel Wiring

$T_{est} = \_\_\_\_$ m
$T_{act} = \_\_\_\_$ m

Section 20.8 Circuits Wired Partially in Series and Partially in Parallel

$T_{est} = \_\_\_\_$ m
$T_{act} = \_\_\_\_$ m

Section 20.9 Internal Resistance

$T_{est} = \_\_\_\_$ m
$T_{act} = \_\_\_\_$ m

Section 20.10 Kirchhoff's Rules

$T_{est} = \_\_\_\_$ m
$T_{act} = \_\_\_\_$ m

Section 20.11 The Measurement of Current and Voltage

T_{est} = _____ m
T_{act} = _____ m

Section 20.12 Capacitors in Series and Parallel

T_{est} = _____ m
T_{act} = _____ m

Section 20.13 RC Circuits

T_{est} = _____ m
T_{act} = _____ m

Section 20.14 Safety and the Physiological Effects of Current

T_{est} = _____ m
T_{act} = _____ m

Section 20.15 Concepts and Calculations

T_{est} = _____ m
T_{act} = _____ m

Additional Notes and Ideas — Chapter 20

21 Magnetic Forces and Magnetic Fields

Teaching Objectives

- Discuss permanent magnets and their properties. State that magnets are always found with two poles that are labeled north and south. Two poles that are the same repel each other while two unlike poles attract each other.

- Introduce the magnetic field and point out its similarity to an electric field. The direction of the magnetic field at each point is the direction that a test magnet, a tiny compass, points. The north pole is indicated by the pole labeled north on the tiny compass. Express the magnitude of the magnetic field as $B = F/(v \sin \theta)$ and give the SI unit for magnetic field, the tesla (T). Make sure students know how to use Right-Hand Rule No. 1 (RHR-1, Section 21.2). Point out that the force on a moving negative charge is opposite to that on a positive charge.
- Discuss the motion of a charged object in a uniform magnetic field. Point out that the constant magnetic force does no work on the charged object because the force is directed perpendicular to its direction of motion. The force does, however, change its direction.
- Explain and demonstrate the effect of a magnetic field on an electric current. Note that RHR-1 can be used to find the direction of the force on the current. Discuss the effect of magnetic forces on a coil or loop of wire carrying a current and provide the formula for the torque, $\tau = NIAB \sin \phi$ where ϕ is the angle between the direction of the magnetic field and the normal to the plane of the loop or coil. Note that the torque is a restoring torque.
- Introduce the concept and demonstrate that electric currents produce magnetic fields. Provide the formulas for magnetic fields produced by currents in long, straight wires; flat, circular coils; and solenoids. Make sure students know how to use Right-Hand Rule No. 2 (RHR-2, Section 21.7). Define and discuss Ampere's law. Provide examples for using Ampere's law.
- Discuss ferromagnetic materials.

Concepts at a Glance

In Section 21.2, the **magnetic force**, like other forces we have encountered (Figures 4.10, 10.12, 11.6, 14.8, and 18.4), may contribute to the **net force** that acts on an object. According to **Newton's second law of motion** (Section 4.3), the **acceleration** is directly proportional to the **net force**.

In Section 21.2, the **electric field** is defined using the **electrostatic force** (Sections 18.2 and 18.5) and a **test charge** (Section 18.6). Similarly, the **magnetic field** is defined in terms of the **magnetic force** and a moving **test charge**.

In Section 21.8, as a continuation of Figure 21.9, **electric currents** (Section 20.1) produce **magnetic fields** (Section 21.2); and these two concepts are related by **Ampere's law**.

List of Transparency Acetates

Figure	Use (✓)	Notes
21.4		
21.7		
21.8		
21.10		
21.12		
21.18		
21.21		
21.24		
21.25		
21.30		
21.39		
21.41		

Lecture Demonstrations

Carpenter and Minnix: B-015, 020, 035, 060, and 115
Freier and Anderson: Ei-1 through 15, 18, 19, and 20, Ej-1, Ej-2, Ep-8, Ep-11, and Er-1, 4 through 9
Hilton: E-6a through 6d, 7a through 7g, and 9a through 9c
Meiners: 31-1.1, 1.8, 1.17, 1.19, 1.20, 1.25, 1.27, and 32-1.1

Laboratory Exercises

#24 Magnetic Field around a Wire
#25 Magnetic Field of a Solenoid

Section 21.1 Magnetic Fields

T_{est} = ____m
T_{act} = ____m

Section 21.2 The Force that a Magnetic Field Exerts on a Moving Charge

T_{est} = ____m
T_{act} = ____m

Lecture Planning and Notes 157

Section 21.3 The Motion of a Charged Particle in a Magnetic Field

$T_{est} = \_\_\_\_ m$
$T_{act} = \_\_\_\_ m$

Section 21.4 The Mass Spectrometer

$T_{est} = \_\_\_\_ m$
$T_{act} = \_\_\_\_ m$

Section 21.5 The Force on a Current in a Magnetic Field

$T_{est} = \_\_\_\_ m$
$T_{act} = \_\_\_\_ m$

Section 21.6 The Torque on a Current-Carrying Coil

$T_{est} = \_\_\_\_ m$
$T_{act} = \_\_\_\_ m$

Section 21.7 Magnetic Fields Produced by Currents

$T_{est} = \_\_\_\_ m$
$T_{act} = \_\_\_\_ m$

Section 21.8 Ampere's Law

$T_{est} = \_\_\_\_ m$
$T_{act} = \_\_\_\_ m$

Section 21.9 Magnetic Materials

T_{est} = _____ m
T_{act} = _____ m

Section 21.10 Concepts & Calculations

T_{est} = _____ m
T_{act} = _____ m

Additional Notes and Ideas — Chapter 21

Lecture Planning and Notes 159

22. Electromagnetic Induction

Teaching Objectives

- Discuss induced emf and induced current and illustrate ways of generating them. Define electromagnetic induction as the process of inducing an emf using a magnetic field. Later, provide the more specific definition in terms of a changing magnetic flux.
- Introduce motional emf and illustrate ways of generating motional emfs.
- Define magnetic flux and point out the similarity to electric flux. Give its SI unit, the weber (Wb). Point out that the flux may be thought of as the number of flux lines that pass through the plane and the importance of the angle that the magnetic field makes with the normal to the plane through which the flux passes.
- State Faraday's law of electromagnetic induction. Provide several examples of correct usage of Faraday's law. Discuss the polarity of the induced emf and define Lenz's law.
- Describe applications of electromagnetic induction including the electric generator. Point out that the current in an electric motor depends on both the applied voltage and the back emf developed due to the rotating coil of the motor.
- Discuss inductors, mutual, and self-induction. Show the symbol for an inductor in a circuit. Introduce the SI unit for inductance, the henry (H). Describe energy storage in the magnetic field of inductors. Define the magnetic energy density in non-conducting regions. Show how transformers work and describe their application.

Concepts at a Glance

In Section 22.3, the concept of **magnetic flux** incorporates both the idea of a **magnetic field** (Section 21.2) and the **surface** through which it passes. This is similar to the approach taken for the **electric flux** in Figure 18.33 (Section 18.9).

In Section 22.4, in extension Figure 22.8, **Faraday's law of electromagnetic induction** specifies the emf created by a change in the **magnetic flux** (Section 22.3) that occurs as time passes. The change in the flux and the **time interval** over which it occurs both appear in Faraday's law.

List of Transparency Acetates

Figure	Use (✓)	Notes
22.1		
22.3		
22.5		
22.10		
22.16		
22.22		
22.26		
22.27		
22.30		

Lecture Demonstrations
   Carpenter and Minnix: B-205, 210, 230, 240, 250, 280, 285, 290, 310, 315 and 320
   Freier and Anderson: Eh-1, 2; Ek-3 through 7; El-1 through 6; Em-1, 2, 4, 5, 8, 10, 12, 13; En-5, 6, 7; Eo-11; Ep-2; Eq-1, 4, 5, 7; and Er-1
   Hilton: E-8a, 8b, 8c, 8d, 11a, 11b, 11c, 11e, and 12a through 12d
   Meiners: 31-2.1, 2.2, 2.6, 2.7, 2.9, 2.15, 3.2, and 3.6

Laboratory Exercises
   #26   Faraday's Law

Section 22.1 Induced EMF and Induced Current

T_{est} = _____ m
T_{act} = _____ m

Section 22.2 Motional EMF

T_{est} = _____ m
T_{act} = _____ m

Section 22.3 Magnetic Flux

T_{est} = ____m
T_{act} = ____m

Section 22.4 Faraday's Law of Electromagnetic Induction

T_{est} = ____m
T_{act} = ____m

Section 22.5 Lenz's Law

T_{est} = ____m
T_{act} = ____m

Section 22.6 Applications of Electromagnetic Induction to the Reproduction of Sound

T_{est} = ____m
T_{act} = ____m

Section 22.7 The Electric Generator

T_{est} = ____m
T_{act} = ____m

Section 22.8 Mutual Inductance and Self-Inductance

T_{est} = ____m
T_{act} = ____m

Section 22.9 Transformers

T_{est} = _____ m
T_{act} = _____ m

Section 22.10 Concepts & Calculations

T_{est} = _____ m
T_{act} = _____ m

Additional Notes and Ideas — Chapter 22

23 Alternating Current Circuits

Teaching Objectives

- Briefly review how a capacitor behaves when it's connected to a dc voltage source. Then, switch the power supply to an alternating voltage source and discuss the behavior of the capacitor. Define the capacitive reactance and remind students of Ohm's law. Discuss the concept of phase and illustrate the phase difference between the current in and voltage across the capacitor. Note that the capacitor, on average, has zero power.
- Briefly review the behavior of inductors. Introduce the concept of inductive reactance. Point out the difference in phase for the current in and the voltage across the inductor when its connected to an ac voltage source. Note that it also consumes no power on average in its operation in an ac circuit.
- Define impedance and provide several examples of *RCL* circuits for calculating impedance and the phase angle. Discuss resonance in *RCL* circuits.
- Introduce the various types of semiconductors and discuss examples of their application.

Concepts at a Glance

In Section 21.4, the phenomenon of **resonance** can occur when an object **vibrates** on a spring (Section 10.8) or when **standing waves** are established on a string (Section 17.5) or in a tube of air (Section 17.6). Resonance can also occur in a **series *RCL* circuit**.

List of Transparency Acetates

Figure	Use (✓)	Notes
23.10		
23.14		
23.20		
23.21		
23.23		
23.30		

164 INSTRUCTOR'S RESOURCE GUIDE — Chapter 5

Lecture Demonstrations
Carpenter and Minnix: B-415,
Freier and Anderson: En-1 through 5; Eo-9 and 13
Hilton: E-13a, 13b, 13c, and 13e
Meiners: 33-2.5 and 2.6

Laboratory Exercises
There are no exercises for this material.

Section 23.1 Capacitors and Capacitive Reactance

T_{est} = ____ m
T_{act} = ____ m

Section 23.2 Inductors and Inductive Reactance

T_{est} = ____ m
T_{act} = ____ m

Section 23.3 Circuits Containing Resistance, Capacitance, and Inductance

T_{est} = ____ m
T_{act} = ____ m

Section 23.4 Resonance in Electric Circuits

T_{est} = ____ m
T_{act} = ____ m

Section 23.5 Semiconductor Devices

T_{est} = ____ m
T_{act} = ____ m

Section 23.6 Concepts & Calculations

$T_{est} = $ _____ m
$T_{act} = $ _____ m

Additional Notes and Ideas   Chapter 23

24 Electromagnetic Waves

Teaching Objectives

- Explain that an electromagnetic wave is composed of electric and magnetic fields that can oscillate either in phase or out of phase. Describe how electromagnetic waves are produced.
- Remind students of the concepts of frequency, wavelength, and speed of waves. Note that electromagnetic waves all have the same speed in a vacuum, $c = 2.998 \times 10^8$ m/s. Point out that electromagnetic waves, unlike sound, do not require a medium and that they travel more slowly through media.
- Discuss the electromagnetic (EM) spectrum, pointing out each region. Help students understand the differences in the wavelengths and frequencies between the various types of EM waves. Students will remember the order of the colors in the visible part of the spectrum if you mention the ROY G. BIV pneumonic (Most students have encountered this before, but they may need a reminder.). Point out that there are far more than just seven colors in the visible spectrum, however.
- Discuss the energy carried by electromagnetic waves and define the total energy density. Show how to calculate the intensity of the EM wave and point out the difference between intensity and power.
- Introduce the Doppler effect for electromagnetic waves and provide examples of using the formula for different source and observer combinations. Note that u is the *relative speed* of the source and the observer and is assumed to be small compared to the speed of light. Compare and contrast the classical expression used for sound waves in Section 16.10. Make sure students understand how to choose the correct sign in the expressions.
- Introduce the linear polarization of light. Point out that the polarization direction is determined by the (fixed) direction in which oscillations of the electric field occur. Describe unpolarized light. Illustrate the use of polarizing materials and discuss their application. Introduce Malus' law and provide examples of its usage.

Concepts at a Glance

In Section 24.1, the **electric field** (Section 18.6) and the **magnetic field** (Sections 21.1 and 21.2) fluctuate together to form an electromagnetic wave.

Lecture Planning and Notes 167

In Section 24.4, the **electromagnetic wave intensity** is defined in similar manner as the **sound intensity** was in Figure 16.22 (Section 16.7). In the case of electromagnetic waves, the intensity is formulated from the concept of **electromagnetic wave power** and the **area** through which the power passes.

List of Transparency Acetates

Figure	Use (✓)	Notes
24.4		
24.10		
24.16		
24.18		
24.20		
24.21		

Lecture Demonstrations

Freier and Anderson: Ep-4, 5, 12, 13; Oa-4; Om-1, 2, 7 through 11, 14 through 19; and On-2

Hilton: O8-a, b, and c

Meiners: 35-6.2 and 6.4

Laboratory Exercises

#27   Polarization

Section 24.1 The Nature of Electromagnetic Waves

$T_{est} = \underline{}$ m
$T_{act} = \underline{}$ m

Section 24.2 The Electromagnetic Spectrum

$T_{est} = \underline{}$ m
$T_{act} = \underline{}$ m

Section 24.3 The Speed of Light

T_{est} = _____ m
T_{act} = _____ m

Section 24.4 The Energy Carried by Electromagnetic Waves

T_{est} = _____ m
T_{act} = _____ m

Section 24.5 The Doppler Effect and Electromagnetic Waves

T_{est} = _____ m
T_{act} = _____ m

Section 24.6 Polarization

T_{est} = _____ m
T_{act} = _____ m

Section 24.7 Concepts & Calculations

T_{est} = _____ m
T_{act} = _____ m

*A*dditional Notes and Ideas   Chapter 24

___

25 The Reflection of Light: Mirrors

Teaching Objectives
- Introduce the concepts of wave front, plane wave, and ray. Point out that the ray gives the direction of travel of the wave.
- State the law of reflection; and discuss specular and diffuse reflection.
- Introduce plane mirrors by placing an object in front of a plane mirror. Show that the reflected rays appear to emanate from behind the mirror. Show that both the image and object lie on the same line that is normal to the plane. Point out that the object and image are both the same distance from the mirror. Define the image as virtual by noting that no light actually comes from the image and point out that the image is always upright and the same size as the object.

- Introduce spherical mirrors and identify the principle axis, the focal point, the radius of curvature, and the center of curvature for concave and convex mirrors. Define paraxial rays. Discuss applications of spherical mirrors.
- Introduce the graphical method of ray tracing using three rays to locate the image for concave (Figure 25.18) and convex (Figure 25.22) mirrors. Stress the importance of using ray diagrams and that they must be constructed carefully. Give examples of drawing diagrams for various locations of the object for both types of mirrors.
- Introduce the mirror equation and the magnification equations. Remind students of the correct procedure in working with reciprocals. Make sure students understand the sign convention used for the variables in the mirror and magnification equations. Indicate problem solving should begin with the construction of a carefully drawn ray diagram to aid thinking and to provide a check for the calculation.

Concepts at a Glance

In Section 25.5, **ray tracing** is used to predict the location and size of an image produced by a spherical mirror. Ray tracing is based on the **law of reflection** (section 25.2) and two points associated with a spherical mirror, its **center of curvature** and its **focal point** (Section 25.4).

In Section 25.6, Figure 25.17 is extended to further describe **image formation** by **spherical mirrors** based on the **law of reflection** (Section 25.2). The mirror equation and the magnification equations use a length associated with the mirror, the **focal length**, and the **image and object distances**.

List of Transparency Acetates

Figure	Use (✓)	Notes
25.3		
25.10		
25.18		
25.19		
25.20		
25.22		

Lecture Demonstrations

Carpenter and Minnix: O-100, 105, 115, 120, 155, 160, 165, and 170
Freier and Anderson: Ob-1 through 11 and Oc-1 through 11
Hilton: O-1c through 1f
Meiners: 34-1.1

Laboratory Exercises
There are no exercises for this material.

Section 25.1 Wave Fronts and Rays

$T_{est} = \underline{}$ m
$T_{act} = \underline{}$ m

Section 25.2 The Reflection of Light

$T_{est} = \underline{}$ m
$T_{act} = \underline{}$ m

Section 25.3 The Formation of Images by a Plane Mirror

$T_{est} = \underline{}$ m
$T_{act} = \underline{}$ m

Section 25.4 Spherical Mirrors

$T_{est} = \underline{}$ m
$T_{act} = \underline{}$ m

Section 25.5 The Formation of Images by Spherical Mirrors

$T_{est} = \underline{}$ m
$T_{act} = \underline{}$ m

Section 25.6 The Mirror Equation and the Magnification Equation

$T_{est} = \underline{}$ m
$T_{act} = \underline{}$ m

Section 25.7 Concepts & Calculations

$T_{est} = \underline{}$ m
$T_{act} = \underline{}$ m

Additional Notes and Ideas — Chapter 25

26 The Refraction of Light: Lenses and Optical Instruments

Teaching Objectives

- Discuss the behavior of light striking the boundary between two media when part of the light is reflected and part is transmitted into the second medium. Draw a ray diagram illustrating the indent, transmitted, and reflected rays. Note that the angle of the transmitted ray is different than the angle of incidence. Make sure students understand how the angles are measured. Introduce the index of refraction and state Snell's law of refraction.
- Introduce the concept of apparent depth and provide illustrative examples that help explain the phenomena.
- Discuss total internal reflection and define the critical angle. Make sure students know how to select n_1 and n_2. Demonstrate fiber optics and discuss their application.
- Define the Brewster angle. Point out that n_1 and n_2 are the refractive indices of the incident and refracting media. Demonstrate the phenomena by reflecting a beam of white light from a glass plate. Show that the incident beam is not polarized using a polarizer. Then show that the reflected light is partially polarized.
- Demonstrate and discuss dispersion of light using a prism.
- Discuss how converging and diverging lenses depend on refraction for image formation. On a diagram, label the principle axis, focal point, the focal length. Show ray diagrams for object placed at various locations relative to converging and diverging lenses.
- Introduce the thin lens equation and the magnification equation. Point out that these equations are the same as those for mirrors. Make sure students understand the sign conventions for using these formulas. Provide examples, with ray diagrams, of using these formulas, including some using more than one lens. Point out that the image of the first lens becomes the object of the second lens.
- Discuss the optics of the eye. Remark that the eye is automatically part of most optical systems. Define near and far points, near sightedness and farsightedness, and refractive power and its unit, the diopter. Illustrate how lenses are used to correct vision. Define angular size and magnification.
- Demonstrate and discuss optical instruments including the magnifying glass, microscope, and telescope. Point out whether images and real or virtual, upright or inverted, and reduced or enlarged.

Concepts at a Glance

In Section 26.1, **the index of refraction** of a material incorporates both the **speed of light in a vacuum** (Section 24.3) and the **speed of light in the material**.

In Section 26.2, Figure 26.1 is extended to include four phenomena that are discussed in the chapter: **total internal reflection** (Section 26.3), **polarization** (Section 26.4), **dispersion** (Section 26.5, and **image formation by lenses** (Sections 26.6 through 26.8), as a way of emphasizing the relationship of the phenomena to the **refraction of light** (Sections 26.1 and 26.2).

In Section 26.8, the figure shows that the equations for **thin lenses** are based on **Snell's law of refraction** (Section 26.2).

List of Transparency Acetates

Figure	Use (✓)	Notes
26.2		
26.10		
26.17		
26.18		
26.25		
26.26		
26.27		
26.28		
26.29		
26.33		

Lecture Demonstrations
  Carpenter and Minnix: O-205, 210, 220, 250, 255, 270, 280, 305, 315, 320, 330, 340, 380, 580, 585, and 590
  Freier and Anderson: Oa-2, 3; Od-1 through 7; Oe-1 through 7; Of-1 through 4; Og-1 through 13; Oi-10 through 12; Oj-9 and 10
  Hilton: O-2b, 2d, 2e, 4a, 4c, 5a, 5b, 5c, 5e, and 5f
  Meiners: 34-1.8, 1.10, 1.11, 1.12, 1.16, and 2.1

Laboratory Exercises
  There are no exercises for this material.

Section 26.1 The Index of Refraction

T_{est} = _____ m
T_{act} = _____ m

Section 26.2 Snell's Law and the Refraction of Light

T_{est} = _____ m
T_{act} = _____ m

Section 26.3 Total Internal Reflection

T_{est} = _____ m
T_{act} = _____ m

Section 26.4 Polarization and the Reflection and Refraction of Light

T_{est} = _____ m
T_{act} = _____ m

Section 26.5 The Dispersion of Light: Prisms and Rainbows

T_{est} = _____ m
T_{act} = _____ m

Section 26.6 Lenses

T_{est} = _____ m
T_{act} = _____ m

Section 26.7 The Formation of Images by Lenses

T_{est} = _____ m
T_{act} = _____ m

Section 26.8 The Thin-Lens Equation and the Magnification Equation

T_{est} = _____ m
T_{act} = _____ m

Section 26.9 Lenses in Combination

T_{est} = _____ m
T_{act} = _____ m

Section 26.10 The Human Eye

T_{est} = _____ m
T_{act} = _____ m

Section 26.11 Angular Magnification and the Magnifying Glass

T_{est} = _____ m
T_{act} = _____ m

Section 26.12 The Compound Microscope

T_{est} = _____ m
T_{act} = _____ m

Section 26.13 The Telescope

T_{est} = _____ m
T_{act} = _____ m

Section 26.14 Lens Aberrations

T_{est} = _____ m
T_{act} = _____ m

Section 26.15 Concepts & Calculations

T_{est} = _____ m
T_{act} = _____ m

Additional Notes and Ideas   **Chapter 26**

27 Interference and the Wave Nature of Light

Teaching Objectives

- Review linear position of sound waves and point out that light waves may also be superposed. Point out that light waves can be interfere constructively if they originate from coherent light sources and define a coherent light source. Illustrate the circumstances under which constructive and destructive interference occur.
- Describe Young's double-slit experiment. Draw a ray from each slit to a point on the screen. Point out that the waves are in phase at the slits, but because the light waves travel different paths, they may not be in phase at the point on the screen. Discuss the historical significance of the experiment in demonstrating the wave nature of light. Remark that the intensity at a given point on the screen is proportional to the square of the total electric field at that point, not the sum of the squares of the individual electric fields.
- Introduce the inference of light reflected from thin transparent films. Point out how the phase difference between the two waves occurs; and make sure that students understand that the wavelength in the medium is used to calculate the phase change that occurs as the light travels through the film. Point out the fact that reflected light will experience a phase change only if the light is traveling from a medium with a lower index of refraction than the medium toward which it's traveling. If possible, demonstrate Newton's rings using a laser, a planar sheet of glass, and a plano-convex lens.
- Introduce Huygens' principle and use it to explain the diffraction of light. Demonstrate the phenomenon using a single slit. Discuss the condition under which constructive interference occurs.
- Discuss the resolving power of optical instruments and explain the Rayleigh criterion. Point out that the minimum angle is expressed in radians, not degrees.
- Pass monochromatic light through a diffraction grating and discuss the conditions for constructive and destructive interference. Discuss the use of a grating spectrometer.
- Discuss the application of interference in the storage and retrieval of information from a compact disc.
- Point out the interference of light is not limited to the visual portion of the spectrum. Describe the use of X-ray diffraction to determine crystalline structure and the construction of arrays of radio telescopes called radio interferometers that are used to improve resolution in radio astronomy.

Concepts at a Glance

In Section 27.1, the **principle of linear superposition** explains various interference phenomena for light waves (Sections 27.2 through 27.4).

In Section 27.5, Figure 27.1 is extended to illustrate how the **principle of linear superposition** (Section 27.1) is also used to describe the **diffraction of light waves** (Section 27.5), the **resolving power of optical instruments** (Section 27.6), and the **diffraction grating** (section 27.7).

List of Transparency Acetates

Figure	Use (✓)	Notes
27.2		
27.3		
27.4		
27.5		
27.19		
27.21		
27.28		
27.29		

Lecture Demonstrations

Carpenter and Minnix: O-405, 455, 460, 465, 505, 510, 520, 525, 530, 550, and 555
Freier and Anderson: Ol-2 through 7, 9 through 18, 21, and 23
Hilton: O-2e; O-7c through 7h, and 7j
Meiners: 35-2.1, 2.2, 2.4, 2.7, and 3.1

Laboratory Exercises

#28   Diffraction of Light

Section 27.1 The Principle of Linear Superposition

T_{est} = ____ m
T_{act} = ____ m

Section 27.2 Young's Double-Slit Experiment

T_{est} = _____ m
T_{act} = _____ m

Section 27.3 Thin-Film Interference

T_{est} = _____ m
T_{act} = _____ m

Section 27.4 The Michelson Interferometer

T_{est} = _____ m
T_{act} = _____ m

Section 27.5 Diffraction

T_{est} = _____ m
T_{act} = _____ m

Section 27.6 Resolving Power

T_{est} = _____ m
T_{act} = _____ m

Section 27.7 The Diffraction Grating

T_{est} = _____ m
T_{act} = _____ m

Section 27.8 Compact Discs, Digital Video Discs, and the Use of Interference

T_{est} = _____ m
T_{act} = _____ m

Section 27.9 X-ray Diffraction

T_{est} = _____ m
T_{act} = _____ m

Section 27.10 Concepts & Calculations

T_{est} = _____ m
T_{act} = _____ m

Additional Notes and Ideas   Chapter 27

Additional Notes and Ideas (*continued*)

Special Relativity

Teaching Objectives
- Define the concepts of *event*, *reference frame*, and *inertial reference frame*.
- Discuss, in general, how physics has evolved and the meaning of physical laws as a relationship between measurable variables. Introduce the postulates of special relativity and discuss the modifications that must be made to the classical variables when the speed of an object is an appreciable fraction of the speed of light. Remark that these modifications do not indicate that the classical physics learned in earlier chapters is wrong, they only indicate a limitation to their validity.
- Explain that every event has associated with it four coordinates (x, y, z, and t) and that the coordinate system and the clock used by an observer may be at rest with respect to an observer. The coordinate system can be viewed as moving to another observer. Discuss the synchronization of clocks and define the proper time interval. Introduce time dilation and the time-dilation equation and provide examples for its proper usage. Point out that the proper time interval is always shorter than the dilated time interval. Mention experiments that have measured and confirmed the predicted time-dilation.
- Define the proper length. State the length-contraction formula and point out that the contraction only occurs in the direction parallel to the direction of motion. Provide examples that show that all observers moving with respect to an object measure a length that is less than the length measured in the frame at rest with the object.

- Introduce the relativistic momentum and provide examples illustrating its calculation. Point out that the correction is consistent with the conservation of momentum for collisions occurring at speeds appreciable compared to the speed of light.
- Discuss the equivalence of mass and energy. State the formula for the total energy of an object of mass m moving with a speed v. Define the rest energy and explain how the relativistic kinetic energy is determined. Discuss the speed of light limitation for objects with mass.
- Demonstrate the correct method for the relativistic addition of velocities.

Concepts at a Glance

In Section 28.2, the chart emphasizes that the **speed of a moving object** must be compared with the **speed of light in a vacuum**, c, to determine whether the effects of **special relativity** are measurably important.

List of Transparency Acetates

Figure	Use (✓)	Notes
28.5		
28.6		

Lecture Demonstrations

There are no demonstrations available for this material.

Laboratory Exercises

There are no laboratory exercises available for this material.

Section 28.1 Events and Inertial Reference Frames

$T_{est} = \underline{\quad}$ m
$T_{act} = \underline{\quad}$ m

Section 28.2 The Postulates of Special Relativity

$T_{est} = \underline{\quad}$ m
$T_{act} = \underline{\quad}$ m

Section 28.3 The Relativity of Time: Time Dilation

T_{est} = _____ m
T_{act} = _____ m

Section 28.4 The Relativity of Length: Length Contraction

T_{est} = _____ m
T_{act} = _____ m

Section 28.5 Relativistic Momentum

T_{est} = _____ m
T_{act} = _____ m

Section 28.6 The Equivalence of Mass and Energy

T_{est} = _____ m
T_{act} = _____ m

Section 28.7 The Relativistic Addition of Velocities

T_{est} = _____ m
T_{act} = _____ m

Additional Notes and Ideas — Chapter 28

29 · Particles and Waves

Teaching Objectives

- Point out that in Chapter 27, light was shown to exhibit the wave-like properties of interference and diffraction, but the wave theory of light cannot account for other phenomena in which electromagnetic waves are involved. Discuss the discovery that particles can also exhibit wave-like properties. Describe wave-particle duality in general terms.
- Define a perfect blackbody and discuss its electromagnetic radiation spectrum as a function of its temperature. Describe Planck's analysis of the problem and the quantized energy solution. Define Planck's constant.

- Illustrate and describe the photoelectric effect experiment and point out why the wave-like nature of light cannot account for the effect because the stopping potential is independent of the intensity of the electromagnetic radiation. More electrons may be emitted by more intense light, but they are not more energetic. Explains Einstein's solution to the problem and define the work function. Provide examples of using the relation: $hf = KE_{max} + W_0$.
- Describe the Compton effect experiment and write down the formula for the phonon momentum. Point out that a photon in the photoelectric effect gives up all of its energy to the emitted electron (and no longer exists), but if only part of the photon energy is given to the electron, the photon has a smaller energy and, thus, a longer wavelength. Describe the scattering via the conservation of momentum and give the Compton formula to calculate the change in the photon wavelength. Note that the experiment confirms that the photon momentum is $p = h/\lambda$.
- Restate that electrons, protons, and all other objects have wave-like properties just as light has particle-like properties. Define the de Broglie wavelength and point out that the p in the formula is the relativistic momentum of the particle. Point out that the wave associated with a particle is one of probability, one can only state the probability of finding the particle, not its exact position. Remind students of the experiments that demonstrate the interference and diffraction of particle-waves.
- Given the probabilistic nature of the particle-wave, one may get a different value every time its position or momentum is measured. Discuss this uncertainty and write down the Heisenberg uncertainty principle that allows one to estimate the uncertainty in each of these quantities.

Concepts at a Glance

In Section 29.3, a moving **particle** has **energy** E (Section 6.4) and a **momentum** p (Section 7.1). Similarly, an **electromagnetic wave** is composed of particle-like packets called photons, each of which also has **energy** (Section 29.3) and **momentum** (Section 29.4).

In Section 29.5, Figure 29.3 is extended to indicate that moving **particles** and **waves** both possess **energy**, **momentum**, and a **wavelength**. The wavelength of a particle is known as its **de Broglie wavelength**.

Lecture Planning and Notes 187

List of Transparency Acetates

Figure	Use (✓)	Notes
29.1		
29.4		
29.10		
29.15		

Lecture Demonstrations
Carpenter and Minnix: S-095
Freier and Anderson: MPb-1
Hilton: A-4b, 4c, 5a, 5b, and 13b
Meiners: 38-2.1, 3.1, 3.2, 7.4, and 39-5.1

Laboratory Exercises
#29 Spectral Lines
#30 Photoelectric Effect – Planck's Constant

Section 29.1 The Wave-Particle Duality

T_{est} = ____ m
T_{act} = ____ m

Section 29.2 Blackbody Radiation and Planck's Constant

T_{est} = ____ m
T_{act} = ____ m

Section 29.3 Photons and the Photoelectric Effect

T_{est} = _____ m
T_{act} = _____ m

Section 29.4 The Momentum of a Photon and the Compton Effect

T_{est} = _____ m
T_{act} = _____ m

Section 29.5 The de Broglie Wavelength and the Wave Nature of Matter

T_{est} = _____ m
T_{act} = _____ m

Section 29.6 The Heisenberg Uncertainty Principle

T_{est} = _____ m
T_{act} = _____ m

Section 29.7 Concepts & Calculations

T_{est} = _____ m
T_{act} = _____ m

Additional Notes and Ideas   Chapter 29

30 The Nature of the Atom

Teaching Objectives
- Describe the structure of the nuclear atom and Rutherford's experiment that helped determine it.
- Discuss line spectra in general and how they may be observed experimentally. Introduce the Lyman, Balmer, and Paschen series within the line spectrum of atomic hydrogen.
- Describe how Bohr constructed his model using ideas from both classical and modern physics. Discuss the assumptions and validity of the model for hydrogen and other elements.
- Introduce quantum mechanics in general. Define the principal, orbital, and magnetic quantum numbers. Remind students of the wave nature of the electron and that we can only state the probability of finding it at a particular location. Point out that electrons don't really move in circular orbits around the nucleus.
- State the Pauli exclusion principle and show how it may be used to understand the periodic table.
- Explain how X-rays are produced and discuss the continuous nature of the X-ray energy spectrum, the *Bremstrahlung*. Point out the characteristic peaks in the spectrum that are determined by the metal used as the target and the sharp cut-off wavelength.
- Demonstrate and describe a laser. Define stimulated and spontaneous emission. Give the characteristics of the light emitted from the laser: coherent, directed, monochromatic, and narrow-beam (that may be very sharply focused). Discuss why each characteristic occurs. List several applications where different types of lasers are used.

Concepts at a Glance

In Section 30.3, **Neils Bohr** developed his **model of the hydrogen atom** in which the **electron energy levels** are **quantized** by combining ideas from **classical** and **modern physics**.

In Section 30.3, Figure 30.6 is extended to show how **Neils Bohr** analyzed the **line spectrum of hydrogen** by combining his **model** of the hydrogen atom with **Einstein's concept** of the **photon**.

List of Transparency Acetates

Figure	Use (✓)	Notes
30.4		
30.5		
30.11		
30.13		
30.14		

Lecture Demonstrations
Freier and Anderson: MPc-1
Hilton: A-2c, 2d, 7, 12, and 20a

Laboratory Exercises
There are no exercises for this material.

Section 30.1 Rutherford Scattering and the Nuclear Atom

$T_{est} = $ ____ m
$T_{act} = $ ____ m

Section 30.2 Line Spectra

$T_{est} = $ ____ m
$T_{act} = $ ____ m

Section 30.3 The Bohr Model of the Hydrogen Atom

$T_{est} = $ ____ m
$T_{act} = $ ____ m

Section 30.4 de Broglie's Explanation of Bohr's Assumption about Angular Momentum

T_{est} = _____ m
T_{act} = _____ m

Section 30.5 The Quantum Mechanical Picture of the Hydrogen Atom

T_{est} = _____ m
T_{act} = _____ m

Section 30.6 The Pauli Exclusion Principle and the Periodic Table of the Elements

T_{est} = _____ m
T_{act} = _____ m

Section 30.7 X-rays

T_{est} = _____ m
T_{act} = _____ m

Section 30.8 The Laser

T_{est} = _____ m
T_{act} = _____ m

Section *30.9 Medical Applications of the Laser

T_{est} = _____ m
T_{act} = _____ m

Section *30.10 Holograms

T_{est} = _____ m
T_{act} = _____ m

Section 30.11 Concepts & Calculations

T_{est} = _____ m
T_{act} = _____ m

Additional Notes and Ideas — Chapter 30

31 Nuclear Physics and Radioactivity

Teaching Objectives

- Introduce the constituents of the nucleus and discuss how modern experiments have helped us to understand nuclear structure. Define atomic number and atomic mass number and introduce the notation for specifying a particular nucleus.
- Discuss the strong nuclear force, indicating that it is one of the three fundamental forces of nature. Contrast the strong nuclear force with the electric force and discuss how the competition between these two forces determines nuclear stability.
- Define the mass defect of the nucleus and discuss the curve of binding energy. Introduce the conversions between joules, atomic mass units, and MeV while explaining the meanings of the latter two units.
- Describe the spontaneous disintegration of an unstable nucleus that leads to radioactivity. Explain the decay reactions that release α, β, and γ rays and the transmutation of elements. Provide examples of involving half-life and the determination of the amount of a radioactive substance remaining after a period of time. Provide examples involving radioactive dating.

Concepts at a Glance

In Section 31.3, the **mass** (Section 4.2) and **rest energy** (Section 28.6) of an object are equivalent in the sense that if one increases (or decreases), the other also increases (or decreases). The **binding energy of a nucleus** is the mass of the separated nucleons minus the mass of the intact nucleus. **All masses are expressed in energy units** by the equivalence of mass and energy (Section 28.6).

In Section 31.4, the **laws of the conservation** of **mass/energy** (Section 6.8 and 28.6), **electric charge** (Section 18.2), **linear momentum** (Section 7.2) and **angular momentum** (Section 9.6) are supplemented by the conservation of **nucleon number** that is used in the analysis of **radioactive decay** processes.

List of Transparency Acetates

Figure	Use (✓)	Notes
31.2		
31.6		
31.14		
31.17		

Lecture Demonstrations
    Carpenter and Minnix: S-135 and 140
    Freier and Anderson: MPa-2
    Hilton: A-15, 15b, 15c, 16, and 18
    Meiners: 41-1.1, 1.8, 1.9, 3.5, and 3.6

Laboratory Exercises
    There are no exercises for this material.

Section 31.1 Nuclear Structure

$T_{est} = \_\_\_\_$ m
$T_{act} = \_\_\_\_$ m

Section 31.2 The Strong Nuclear Force and the Stability of the Nucleus

$T_{est} = \_\_\_\_$ m
$T_{act} = \_\_\_\_$ m

Section 31.3 The Mass Defect of the Nucleus and Nuclear Binding Energy

T_{est} = _____ m
T_{act} = _____ m

Section 31.4 Radioactivity

T_{est} = _____ m
T_{act} = _____ m

Section 31.5 The Neutrino

T_{est} = _____ m
T_{act} = _____ m

Section 31.6 Radioactive Decay and Activity

T_{est} = _____ m
T_{act} = _____ m

Section 31.7 Radioactive Dating

T_{est} = _____ m
T_{act} = _____ m

Section 31.8 Radioactive Decay Series

T_{est} = _____ m
T_{act} = _____ m

Section 31.9 Radiation Detectors

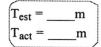
T_{est} = _____ m
T_{act} = _____ m

Section 31.10 Concepts & Calculations

T_{est} = _____ m
T_{act} = _____ m

*A*dditional Notes and Ideas — Chapter 31

32 Ionizing Radiation, Nuclear Energy, and Elementary Particles

Teaching Objectives

- Introduce ionizing radiation and define exposure, absorbed dose, relative biological effectiveness, and biologically equivalent dose along with the units of roentgen, gray, rad, and rem. Demonstrate the meaning of each of these measures of ionizing radiation. Discuss the effects of ionizing radiation on humans.
- Describe induced nuclear reactions and discuss the difference between induced and spontaneous reactions. Illustrate the discussion with some possible reactions. Discuss the fission process and provide examples of fission reactions. Point out how one might generate a chain reaction and show how fission reactions are controlled in nuclear reactors.
- Introduce fusion reactions and provide examples of them. Describe the fusion reactions that occur in stars that generate the elements (up to iron). Discuss efforts underway to construct controlled fusion reactors.
- Introduce the field of particle physics and discuss efforts to unify the fundamental forces via the standard model and the role of the various types of particles. Discuss elementary particles and their classification.
- Define the science of cosmology and point out how the work of particle physicists has helped explain the past, present, and future of the universe. Discuss the expanding universe and the evidence for the Big Bang theory and the development of the standard cosmological model for the evolution of the universe.

Concepts at a Glance

In Section 32.1, Figure 31.7 (Section 31.4) is extended to illustrate that the **conservation laws** are also obeyed in **induced nuclear reactions**.

List of Transparency Acetates

Figure	Use (✓)	Notes
32.3		
32.15		

Lecture Demonstrations
    Freier and Anderson: MPa-1
    Hilton: A-22 and 23
    Meiners: 41-2.9

Laboratory Exercises
    There are no laboratory exercises available for this material.

Section 32.1 Biological Effects of Ionizing Radiation

T_{est} = ____m
T_{act} = ____m

Section 32.2 Induced Nuclear Reactions

T_{est} = ____m
T_{act} = ____m

Section 32.3 Nuclear Fission

T_{est} = ____m
T_{act} = ____m

Section 32.4 Nuclear Reactors

$T_{est} = \underline{} m$
$T_{act} = \underline{} m$

Section 32.5 Nuclear Fusion

$T_{est} = \underline{} m$
$T_{act} = \underline{} m$

Section 32.6 Elementary Particles

$T_{est} = \underline{} m$
$T_{act} = \underline{} m$

Section 32.7 Cosmology

$T_{est} = \underline{} m$
$T_{act} = \underline{} m$

Section 32.8 Concepts & Calculations

$T_{est} = \underline{} m$
$T_{act} = \underline{} m$

Additional Notes and Ideas   Chapter 32

APPENDIX A

College Physics Textbook Comparison and Conversion

The table below provides a listing of topics and the chapter sections in which they appear in competing algebra/trigonometry-based physics textbooks. This table is provided as an aid to instructors who are adopting Cutnell & Johnson's text and had used one of the other texts previously. The standard has become to break up the two semester or three quarter course into the following six sections: mechanics, thermal physics, waves motion & sound, electricity & magnetism, light & optics, and modern physics. This sectioning is used in the table below. Selected topics within each section are listed in the left-hand column. The chapter(s) and section(s) in which the topic is presented is(are) indicated in the column for each text. For example, "14.1,7-9" indicates that the topic is treated in chapter 14 in sections 1, 7, 8, and 9.

A detailed comparison between these books is probably of limited usefulness since the coverage is nearly the same in all of the books. Differences in the texts arise out of the ordering of certain topics based on the personal preferences of the authors, physics education research that suggests the ordering, or the need for a clean break in the material between semesters or quarters.

The table doesn't indicate the depth or breadth of coverage of a given topic. All texts present the basic physics of each topic; then, features are added such as example problems, conceptual examples, recent research results, essays, applications, drawings and photos to promote understanding and interest in physics. Adding too many layers can overwhelm the student; and adding too few can fail to communicate the relevance of the subject, and bore and frustrate the student. There is a need to cover as much of the text as possible, especially for pre-med. students preparing for the MCAT; and therefore, a delicate balance must be drawn between depth and breadth both within a given text and in the presentation of the course itself. Chapter 4 of this guide suggests ways in which the Cutnell & Johnson text can be effectively used in either a two semester of a three quarter format; and Chapter 5 helps you plan your lectures and, when possible, suggests topics which may be omitted in the lecture presentation without loss of continuity.

College Physics Textbook Comparison and Conversion

Mechanics

Topic	Cutnell & Johnson	Giancoli	Serway & Faughn
Introduction & Units	1.1-4	1.1-8	1.1-10
Vectors & Scalars	1.5-9	3.1-4	3.1-3
Kinematics, 1D	2.1-8	2.1-8	2.1-7
Kinematics, 2D	3.1-5	3.5-8	3.4-6
Newton's Laws	4.1-5	4.1-7	4.1-4
Gravitational Force	4.6-7	5.6-9	7.7
Applications of Newton's Laws	4.8-13	4.8	4.5-6
Uniform Circular Motion	5.1-8	5.1-5	7.4-6
Work, Energy, & Power	6.1-10	6.1-10	5.1-8
Impulse & Momentum	7.1-2, 7.6	7.1-3	6.1-2
Collisions	7.3-4	7.4-7	6.3-4
Center of Mass	7.5	7.8-10	
Rotational Kinematics	8.1-8	8.1-3	7.1-3
Rotational Dynamics	9.1-9.7	8.4-9	8.1-7
Harmonic Motion	10.1-6, 10.9	11.1-6	13.4-7
Mechanical Properties	10.7-9	9.6-7	9.2
Fluid Statics	11.1-6, 11.12	10.1-6	9.4-6
Fluid Dynamics	11.7-11.12	10.7-11	9.7

Thermal Physics

Topic	Cutnell & Johnson	Giancoli	Serway & Faughn
Temperature Scales	12.1-3	13.2	10.1-2
Thermal Expansion	12.4-5, 12.11	13.4-6	10.3
Calorimetry and Phase Changes	12.6-9, 12.11	14.5-6	11.3-4
Heat Transfer	13.1-5	14.7-9	11.5-8
Ideal Gases and Kinetic Theory	14.1-5	13.1, 7-13	10.4-6
Thermodynamics: Zeroth and First Laws	15.1-3, 6, 13	13.3, 15.1-3	12.1-3
Thermodynamics: Second and Third Laws	15.4, 5, 7-13	15.4-11	12.4-6

Waves and Sound

Topic	Cutnell & Johnson	Giancoli	Serway & Faughn
Wave Properties	16.1-4	11.7-10	13.8-9
Sound	16.5-8, 10, 11	12.1-7	14.1-5
Doppler Effect (Sound)	16.9, 16.12	12.8-10	14.6
Superposition and Interference	17.1-8	11.11,-13	13.12, 14.7-8

Electricity and Magnetism

Topic	Cutnell & Johnson	Giancoli	Serway & Faughn
Electric Charge, Forces, and Fields	18.1-8, 10, 11	16.1-9	15.1-6
Shielding and Gauss' Law	18.9	App. D	15.10
Electric Potential & Electric Potential Energy	19.1-4, 6, 7	17.1-6	16.1-4
Capacitors & Dielectrics	19.5	17.7-9	16.6-10
EMF & Current	20.1	18.1-2, 19.2	18.1, 17.1-2
Ohm's Law/Resistance	20.2-3	18.3-5	17.3-5
Electric Power	20.4	18.6-7	17.7
ac current & voltage	20.5	18.8	18.6
Electric Circuits	20.6-11, 14	19.1, 3-5	18.1-4
Capacitor Circuits	20.12	19.6-7	18.5
Magnetic Fields and Forces	21.1-10	20.1-15	19.1-12
Electromagnetic Induction	22.1-10	21.1-11	20.1-9
ac Circuits	23.1-6	21.12-15	21.1-7

Light and Optics

Topic	Cutnell & Johnson	Giancoli	Serway & Faughn
Electromagnetic Waves	24.1-7	22.1-8,	22.1-3
Reflection	25.1-6	23.1-3	22.4, 23.1-3
Refraction	26.1-9, 15	23.4-9	22.4-6, 23.4-5
The Eye & Optical Instruments	26.10-14, 15	23.10-11	25.1-7
Interference and the Wave Nature of Light	27.1-10	24.1-11	24.1-7, 25.8

Modern Physics

Topic	Cutnell & Johnson	Giancoli	Serway & Faughn
Special Relativity	28.1-7	26.1-12	26.1-10
Particles & Waves	29.1-7	27.1-6	27.1-3
Atomic Models	30.1-5	27.8-11	28.1-6
Pauli Exclusion Principle and the Periodic Table	30.6, 12	28.7-8	28.9
X-rays	30.7, 12	28.9	27.4, 28.11
Lasers	30.8-10	28.11	28.13
Nuclear Physics	31.1-3, 10	30.1-2	29.1-2
Radioactivity	31.4-10	30.3-13	29.3-6
Ionizing Radiation	32.1, 8	31.4-7	29.7
Nuclear Reactions	32.2-5	31.2-3	30.1-3
Elementary Particles	32.6, 8	32.1-11	30.4-13
Cosmology	32.7	33.1-7	30.15

APPENDIX B

Homework Problem Locator

The tables below allow one to readily find homework problem numbers in the sixth edition (*6e*) of *Physics* relative to the fifth edition (*5e*). Problems in *6e* that have been numerically altered or substantially rewritten from the *5e* problem are indicated by a bracket (for example, {8}); and those that have been deleted from *6e* are indicated by a minus sign (−). Problems that are new in *6e* are also listed.

Note: Indicators are used in the sixth edition of *Physics* for problems that have solutions available via the worldwide web (**www**) or in the *Student's Solutions Manual* (**ssm**). This availability may be taken into consideration when assigning homework problems.

Chapter 1: Introduction and Mathematical Concepts

New Problems in *6e*:			4, 10, 14, 18, 26, 29, 34, 37, 46, 49		
5e	6e	5e	6e	5e	6e
1	5	24	24	47	47
2	2	25	23	48	48
3	3	26	56	49	59
4	−	27	27	50	50
5	1	28	−	51	12
6	6	29	28	52	22
7	7	30	60	53	31
8	54	31	53	54	8
9	9	32	32	55	43
10	−	33	33	56	−
11	13	34	−	57	35
12	−	35	57	58	−
13	52	36	36	59	−
14	11	37	58	60	30
15	15	38	38	61	62
16	−	39	39	62	61
17	16	40	42	63	63
18	17	41	41	64	64
19	19	42	40	65	65
20	20	43	55	66	66
21	21	44	45	67	67
22	51	45	44	68	68
23	25	46	−	69	69

Chapter 2: Kinematics in One Dimension

New Problems in *6e*:			6, 10, 16, 24, 26, 29, 64, 71, 76, 87, 88		
5e	*6e*	*5e*	*6e*	*5e*	*6e*
1	3	30	30	60	60
2	2	31	32	61	61
3	1	32	{31}	62	62
4	69	33	77	63	63
5	66	34	34	64	–
6	5	35	79	65	37
7	7	36	36	66	–
8	74	37	65	67	13
9	9	38	38	68	42
10	–	39	40	69	4
11	11	40	39	70	70
12	12	41	41	71	–
13	67	42	68	72	33
14	14	43	44	73	73
15	15	44	43	74	8
16	–	45	45	75	51
17	{17}	46	46	76	–
18	18	47	{47}	77	72
19	21	48	{49}	78	55
20	20	49	48	79	35
21	19	50	50	80	80
22	22	51	75	81	81
23	25	52	52	82	82
24	23	53	53	83	83
25	–	54	54	84	84
26	–	55	56	85	85
27	27	56	78	86	86
28	28	57	57		
29	–	58	59		

Chapter 3: Kinematics in Two Dimensions

New Problems in *6e*:			8, 18, 26, 32, 42, 48, 54, 58, 67		
5e	*6e*	*5e*	*6e*	*5e*	*6e*
1	5	11	11	21	21
2	60	12	64	22	66
3	3	13	61	23	22
4	4	14	14	24	24
5	1	15	63	25	65
6	7	16	16	26	20
7	–	17	17	27	27
8	6	18	19	28	31
9	9	19	–	29	29
10	10	20	–	30	30

Chapter 3 (continued)

5e	6e	5e	6e	5e	6e
31	28	48	–	65	25
32	–	49	49	66	23
33	37	50	52	67	–
34	34	51	51	68	36
35	{35}	52	50	69	57
36	68	53	53	70	70
37	33	54	–	71	41
38	{39}	55	56	72	72
39	38	56	{55}	73	44
40	40	57	69	74	74
41	71	58	–	75	75
42	–	59	59	76	76
43	43	60	2	77	77
44	73	61	13	78	78
45	45	62	62	79	79
46	46	63	15		
47	47	64	12		

Chapter 4: Forces and Newton's Laws of Motion

New Problems in 6e: 2, 14, 26, 42, 46, 58, 68, 74, 88, 90, 118, 119

5e	6e	5e	6e	5e	6e
1	1	27	23	53	53
2	–	28	28	54	54
3	7	29	31	55	102
4	4	30	30	56	59
5	5	31	29	57	107
6	96	32	106	58	–
7	3	33	33	59	{56}
8	8	34	34	60	60
9	9	35	35	61	111
10	10	36	94	62	62
11	63	37	37	63	12
12	{11}	38	38	64	64
13	13	39	91	65	65
14	15	40	{41}	66	92
15	–	41	40	67	67
16	16	42	–	68	{72}
17	17	43	43	69	69
18	{18}	44	44	70	93
19	21	45	45	71	95
20	20	46	47	72	–
21	19	47	–	73	73
22	98	48	52	74	77
23	87	49	49	75	103
24	24	50	50	76	76
25	25	51	89	77	–
26	–	52	48	78	78

Chapter 4 (continued)

5e	6e	5e	6e	5e	6e
79	79	93	70	107	57
80	80	94	36	108	108
81	105	95	71	109	85
82	82	96	6	110	110
83	83	97	–	111	61
84	84	98	22	112	112
85	109	99	99	113	113
86	86	100	100	114	114
87	27	101	101	115	115
88	–	102	55	116	116
89	51	103	75	117	117
90	97	104	104		
91	39	105	81		
92	66	106	32		

Chapter 5: Dynamics of Uniform Circular Motion

New Problems in 6e:				2, 3, 12, 23, 24, 31, 36, 40	
5e	6e	5e	6e	5e	6e
1	1	22	–	43	11
2	–	23	22	44	28
3	–	24	–	45	7
4	4	25	25	46	46
5	47	26	26	47	5
6	6	27	27	48	48
7	45	28	44	49	49
8	52	29	29	50	30
9	9	30	50	51	51
10	10	31	–	52	8
11	43	32	34	53	55
12	–	33	33	54	54
13	15	34	32	55	53
14	14	35	35	56	56
15	13	36	–	57	57
16	16	37	39	58	59
17	17	38	38	59	58
18	{19}	39	37	60	60
19	18	40	{41}	61	61
20	20	41	–		
21	21	42	42		

Chapter 6: Work and Energy

New Problems in *6e*:				4, 12, 20, 30, 33, 40, 52, 55, 64	
5e	*6e*	*5e*	*6e*	*5e*	*6e*
1	71	30	–	59	74
2	–	31	31	60	60
3	3	32	32	61	61
4	2	33	–	62	62
5	5	34	34	63	65
6	6	35	69	64	–
7	7	36	36	65	63
8	73	37	37	66	66
9	9	38	–	67	67
10	{10}	39	77	68	68
11	11	40	{39}	69	35
12	14	41	41	70	70
13	13	42	42	71	1
14	–	43	43	72	47
15	17	44	78	73	8
16	16	45	45	74	59
17	15	46	46	75	49
18	18	47	72	76	{75}
19	19	48	48	77	38
20	22	49	50	78	44
21	23	50	76	79	79
22	–	51	–	80	80
23	21	52	{51}	81	81
24	24	53	53	82	82
25	27	54	54	83	83
26	26	55	–	84	84
27	25	56	56	85	85
28	28	57	57		
29	29	58	58		

Chapter 7: Impulse and Momentum

New Problems in *6e*:				3, 4, 10, 16, 20, 26, 28, 36, 56, 64, 65	
5e	*6e*	*5e*	*6e*	*5e*	*6e*
1	5	13	11	25	25
2	48	14	14	26	49
3	2	15	15	27	27
4	–	16	46	28	–
5	1	17	47	29	29
6	50	18	18	30	32
7	7	19	22	31	31
8	8	20	–	32	30
9	9	21	53	33	{55}
10	–	22	{19}	34	–
11	{13}	23	57	35	35
12	12	24	24	36	34

Chapter 7 (continued)

5e	6e	5e	6e	5e	6e
37	37	46	—	55	33
38	40	47	17	56	44
39	39	48	—	57	23
40	38	49	—	58	58
41	41	50	6	59	59
42	52	51	51	60	60
43	43	52	42	61	61
44	—	53	21	62	62
45	45	54	54	63	63

Chapter 8: Rotational Kinematics

New Problems in *6e*: 2, 6, 8, 16, 21, 25, 30, 44, 49, 52, 61, 75, 76

5e	6e	5e	6e	5e	6e
1	3	26	26	51	63
2	—	27	27	52	53
3	1	28	28	53	54
4	4	29	32	54	—
5	5	30	—	55	55
6	—	31	29	56	—
7	7	32	33	57	39
8	—	33	—	58	{31}
9	9	34	34	59	17
10	10	35	35	60	60
11	11	36	36	61	58
12	64	37	68	62	62
13	13	38	38	63	51
14	14	39	57	64	12
15	15	40	40	65	65
16	—	41	41	66	42
17	59	42	66	67	67
18	18	43	43	68	37
19	19	44	—	69	69
20	—	45	45	70	70
21	20	46	48	71	71
22	22	47	47	72	72
23	23	48	46	73	73
24	—	49	56	74	74
25	24	50	50		

Chapter 9: Rotational Dynamics

New Problems in 6e:			3, 8, 12, 14, 30, 34, 36, 46, 54, 66, 79, 80		
5e	6e	5e	6e	5e	6e
1	61	27	26	53	53
2	2	28	28	54	52
3	4	29	29	55	63
4	6	30	–	56	56
5	5	31	31	57	57
6	–	32	32	58	69
7	–	33	65	59	60
8	7	34	35	60	59
9	9	35	62	61	1
10	10	36	–	62	–
11	–	37	39	63	–
12	11	38	38	64	{55}
13	13	39	37	65	33
14	12	40	40	66	18
15	15	41	41	67	21
16	16	42	72	68	24
17	{17}	43	45	69	58
18	64	44	–	70	48
19	19	45	43	71	25
20	–	46	44	72	42
21	67	47	70	73	73
22	22	48	47	74	74
23	23	49	49	75	75
24	68	50	50	76	76
25	71	51	51	77	77
26	27	52	–	78	78

Chapter 10: Simple Harmonic Motion and Elasticity

New Problems in 6e:			6, 14, 19, 28, 30, 44, 54, 62, 75, 76, 77, 91, 92		
5e	6e	5e	6e	5e	6e
1	1	15	15	29	25
2	72	16	16	30	{31}
3	3	17	17	31	–
4	4	18	70	32	32
5	5	19	67	33	33
6	–	20	20	34	{35}
7	–	21	21	35	–
8	8	22	74	36	80
9	10	23	23	37	{37}
10	{9}	24	24	38	38
11	7	25	29	39	69
12	12	26	27	40	42
13	13	27	26	41	41
14	–	28	–	42	40

Chapter 10 (continued)

5e	6e	5e	6e	5e	6e
43	43	59	59	75	11
44	–	60	60	76	58
45	45	61	61	77	–
46	46	62	–	78	79
47	47	63	63	79	81
48	73	64	64	80	36
49	71	65	65	81	82
50	52	66	84	82	34
51	51	67	–	83	83
52	50	68	68	84	66
53	53	69	39	85	85
54	56	70	18	86	86
55	57	71	49	87	87
56	–	72	2	88	88
57	55	73	48	89	89
58	78	74	22	90	90

Chapter 11: Fluids

New Problems in 6e: 4, 8, 10, 14, 20, 24, 33, 38, 46, 52, 58, 60, 63, 73, 86, 102, 103

5e	6e	5e	6e	5e	6e
1	5	28	28	55	55
2	82	29	90	56	56
3	–	30	30	57	57
4	81	31	31	58	–
5	1	32	–	59	59
6	7	33	32	60	–
7	6	34	34	61	61
8	–	35	35	62	93
9	9	36	37	63	–
10	11	37	36	64	67
11	–	38	–	65	65
12	12	39	79	66	66
13	13	40	–	67	64
14	–	41	42	68	96
15	{15}	42	41	69	69
16	88	43	43	70	–
17	17	44	44	71	71
18	95	45	45	72	70
19	19	46	–	73	72
20	77	47	47	74	84
21	21	48	94	75	75
22	22	49	48	76	76
23	85	50	50	77	–
24	25	51	83	78	78
25	80	52	53	79	39
26	26	53	–	80	–
27	91	54	87	81	3

Chapter 11 (continued)

5e	6e	5e	6e	5e	6e
82	2	90	29	98	98
83	51	91	27	99	99
84	74	92	92	100	100
85	16	93	62	101	101
86	40	94	49	102	104
87	23	95	18		
88	54	96	68		
89	89	97	97		

Chapter 12: Temperature and Heat

New Problems in 6e: 2, 14, 16, 20, 27, 28, 34, 40, 58, 64, 68, 70, 101, 102

5e	6e	5e	6e	5e	6e
1	1	35	—	69	72
2	4	36	36	70	76
3	3	37	37	71	71
4	—	38	38	72	—
5	5	39	—	73	69
6	86	40	39	74	74
7	7	41	45	75	73
8	6	42	42	76	75
9	9	43	{43}	77	10
10	77	44	44	78	52
11	11	45	41	79	55
12	12	46	47	80	80
13	81	47	46	81	13
14	—	48	48	82	82
15	15	49	91	83	31
16	—	50	50	84	84
17	17	51	51	85	85
18	18	52	78	86	8
19	19	53	53	87	61
20	90	54	54	88	59
21	—	55	79	89	{89}
22	22	56	56	90	31
23	23	57	57	91	49
24	94	58	—	92	92
25	25	59	88	93	93
26	26	60	60	94	24
27	29	61	87	95	95
28	—	62	62	96	96
29	—	63	{63}	97	97
30	30	64	—	98	98
31	83	65	65	99	99
32	32	66	66	100	100
33	35	67	67		
34	33	68	—		

Chapter 13: The Transfer of Heat

New Problems in *6e*:				4, 12, 16, 20, 29, 43, 44	
5e	*6e*	*5e*	*6e*	*5e*	*6e*
1	27	15	35	29	—
2	28	16	18	30	22
3	—	17	19	31	21
4	3	18	—	32	{13}
5	5	19	17	33	24
6	8	20	—	34	10
7	7	21	31	35	15
8	6	22	30	36	14
9	—	23	23	37	37
10	34	24	32	38	38
11	11	25	25	39	39
12	9	26	26	40	40
13	{33}	27	1	41	41
14	36	28	2	42	42

Chapter 14: The Ideal Gas Law and Kinetic Theory

New Problems in *6e*:				4, 12, 20, 21, 32, 40, 46, 51, 63, 64	
5e	*6e*	*5e*	*6e*	*5e*	*6e*
1	5	22	22	43	41
2	—	23	23	44	44
3	3	24	24	45	45
4	2	25	57	46	—
5	1	26	26	47	11
6	6	27	27	48	13
7	55	28	50	49	43
8	8	29	31	50	—
9	9	30	30	51	28
10	10	31	29	52	16
11	47	32	—	53	53
12	—	33	33	54	36
13	48	34	34	55	7
14	14	35	{35}	56	56
15	—	36	54	57	25
16	52	37	37	58	58
17	17	38	38	59	59
18	18	39	39	60	60
19	19	40	—	61	61
20	—	41	49	62	62
21	—	42	42		

Chapter 15: Thermodynamics

New Problems in 6e:		4, 8, 18, 21, 26, 30, 44, 48, 50, 60, 66, 85, 99, 100			
5e	*6e*	*5e*	*6e*	*5e*	*6e*
1	5	34	80	67	67
2	2	35	35	68	68
3	–	36	89	69	{69}
4	3	37	37	70	70
5	1	38	88	71	79
6	6	39	39	72	72
7	–	40	40	73	{73}
8	77	41	41	74	20
9	9	42	76	75	19
10	12	43	43	76	42
11	11	44	–	77	7
12	10	45	91	78	57
13	83	46	47	79	71
14	84	47	46	80	34
15	15	48	–	81	61
16	16	49	49	82	51
17	17	50	–	83	13
18	–	51	82	84	14
19	75	52	52	85	86
20	74	53	53	86	–
21	–	54	90	87	87
22	22	55	55	88	38
23	23	56	56	89	36
24	24	57	78	90	54
25	{27}	58	58	91	45
26	–	59	59	92	28
27	25	60	–	93	93
28	92	61	81	94	94
29	29	62	62	95	95
30	32	63	–	96	96
31	31	64	64	97	97
32	–	65	65	98	98
33	33	66	63		

Chapter 16: Waves and Sound

New Problems in 6e:		2, 8, 30, 38, 42, 58, 65, 70, 74, 77, 84, 109, 110			
5e	*6e*	*5e*	*6e*	*5e*	*6e*
1	1	8	–	15	85
2	–	9	9	16	16
3	5	10	96	17	{17}
4	4	11	11	18	18
5	3	12	{12}	19	97
6	86	13	13	20	20
7	7	14	14	21	21

Chapter 16 (continued)

5e	6e	5e	6e	5e	6e
22	{22}	51	88	80	80
23	23	52	52	81	81
24	24	53	49	82	–
25	90	54	54	83	33
26	26	55	{55}	84	15
27	27	56	56	85	–
28	28	57	57	86	6
29	29	58	–	87	61
30	82	59	59	88	51
31	89	60	60	89	31
32	32	61	87	90	25
33	83	62	62	91	73
34	92	63	93	92	34
35	35	64	63	93	94
36	37	65	–	94	95
37	36	66	100	95	19
38	–	67	67	96	10
39	39	68	68	97	98
40	40	69	69	98	99
41	41	70	–	99	64
42	44	71	71	100	66
43	43	72	72	101	45
44	–	73	91	102	102
45	101	74	–	103	103
46	46	75	75	104	104
47	47	76	–	105	105
48	48	77	76	106	106
49	53	78	78	107	107
50	50	79	79	108	108

Chapter 17: The Principle of Linear Superposition and Interference Phenomena

New Problems in *6e*: 2, 8, 10, 18, 24, 29, 34, 38, 50, 52, 53, 62, 63

5e	6e	5e	6e	5e	6e
1	3	14	15	27	51
2	–	15	14	28	26
3	1	16	–	29	–
4	6	17	19	30	30
5	47	18	16	31	31
6	4	19	17	32	32
7	7	20	48	33	33
8	–	21	21	34	36
9	9	22	54	35	–
10	12	23	23	36	46
11	11	24	45	37	37
12	–	25	25	38	–
13	13	26	28	39	39

Chapter 17 (continued)

5e	6e	5e	6e	5e	6e
40	42	48	20	56	56
41	40	49	{49}	57	57
42	41	50	–	58	58
43	43	51	27	59	59
44	44	52	–	60	60
45	–	53	–	61	61
46	35	54	22		
47	5	55	55		

Chapter 18: Electric Forces and Electric Fields

New Problems in *6e*:   3, 15, 16, 20, 28, 36, 53, 58, 65, 75, 76

5e	6e	5e	6e	5e	6e
1	1	25	25	49	49
2	–	26	26	50	50
3	2	27	59	51	51
4	4	28	29	52	52
5	5	29	30	53	54
6	6	30	–	54	55
7	56	31	31	55	7
8	8	32	32	56	57
9	9	33	–	57	60
10	10	34	35	58	–
11	61	35	33	59	27
12	14	36	34	60	62
13	–	37	37	61	11
14	12	38	38	62	63
15	13	39	39	63	–
16	17	40	66	64	64
17	–	41	41	65	19
18	18	42	42	66	40
19	67	43	45	67	24
20	–	44	44	68	69
21	21	45	43	69	70
22	22	46	48	70	71
23	23	47	47	71	72
24	68	48	46	72	73
				73	74

Chapter 19: Electric Potential Energy and the Electric Potential

New Problems in 6e:		4, 12, 14, 16, 28, 32, 42, 54, 62, 67, 68			
5e	*6e*	*5e*	*6e*	*5e*	*6e*
1	1	23	19	45	44
2	50	24	60	46	46
3	–	25	25	47	59
4	3	26	58	48	48
5	5	27	27	49	–
6	52	28	–	50	2
7	{7}	29	29	51	39
8	8	30	30	52	6
9	9	31	55	53	38
10	10	32	–	54	–
11	11	33	{33}	55	31
12	49	34	56	56	34
13	13	35	35	57	{57}
14	–	36	36	58	26
15	15	37	37	59	47
16	–	38	53	60	24
17	17	39	51	61	61
18	18	40	40	62	62
19	23	41	{41}	63	63
20	20	42	–	64	64
21	{21}	43	43	65	65
22	22	44	45	66	66

Chapter 20: Electric Circuits

New Problems in 6e:		1, 8, 12, 16, 26, 30, 54, 68, 82, 86, 123, 124			
5e	*6e*	*5e*	*6e*	*5e*	*6e*
1	–	20	20	39	43
2	2	21	100	40	40
3	3	22	22	41	41
4	–	23	23	42	42
5	103	24	24	43	39
6	4	25	–	44	108
7	7	26	25	45	{45}
8	6	27	27	46	46
9	9	28	28	47	47
10	10	29	29	48	48
11	11	30	32	49	51
12	14	31	31	50	50
13	13	32	36	51	53
14	–	33	33	52	52
15	105	34	34	53	{49}
16	–	35	35	54	–
17	17	36	–	55	55
18	113	37	37	56	56
19	19	38	38	57	57

Chapter 20 (continued)

5e	6e	5e	6e	5e	6e
58	60	80	80	102	102
59	59	81	81	103	5
60	{58}	82	107	104	76
61	61	83	–	105	15
62	62	84	84	106	106
63	{64}	85	85	107	83
64	63	86	–	108	114
65	109	87	87	109	65
66	66	88	88	110	110
67	–	89	101	111	111
68	67	90	90	112	72
69	69	91	91	113	18
70	71	92	92	114	44
71	70	93	115	115	93
72	112	94	94	116	116
73	73	95	97	117	117
74	74	96	96	118	118
75	75	97	95	119	119
76	104	98	{99}	120	120
77	77	99	{98}	121	121
78	{78}	100	21	122	122
79	79	101	89		

Chapter 21: Magnetic Forces and Magnetic Fields

New Problems in 6e:	6, 12, 17, 22, 24, 26, 32, 36, 40, 50, 75, 83, 85				
5e	6e	5e	6e	5e	6e
1	69	23	73	45	77
2	2	24	–	46	46
3	64	25	25	47	47
4	4	26	28	48	48
5	5	27	27	49	65
6	66	28	–	50	68
7	7	29	29	51	51
8	–	30	30	52	52
9	9	31	31	53	72
10	11	32	70	54	54
11	–	33	33	55	{55}
12	10	34	{34}	56	74
13	15	35	35	57	57
14	14	36	67	58	76
15	12	37	37	59	59
16	16	38	38	60	60
17	71	39	–	61	61
18	18	40	42	62	62
19	19	41	41	63	63
20	20	42	39	64	3
21	{21}	43	{43}	65	49
22	–	44	44	66	–

Chapter 21 (continued)

5e	6e	5e	6e	5e	6e
67	–	73	23	80	80
68	–	74	56	81	81
69	1	75	8	82	82
70	–	76	58	83	84
71	–	78	45		
72	53	79	78		

Chapter 22: Electromagnetic Induction

New Problems in 6e: 2, 6, 10, 18, 20, 40, 48, 60, 81, 82

5e	6e	5e	6e	5e	6e
1	1	28	{28}	55	57
2	–	29	65	56	56
3	3	30	30	57	{55}
4	4	31	31	58	58
5	5	32	70	59	59
6	–	33	{33}	60	–
7	7	34	34	61	61
8	8	35	39	62	62
9	9	36	37	63	{63}
10	–	37	–	64	24
11	11	38	36	65	29
12	14	39	67	66	66
13	13	40	38	67	35
14	12	41	41	68	68
15	15	42	42	69	{69}
16	16	43	74	70	32
17	–	44	44	71	25
18	–	45	49	72	72
19	19	46	46	73	43
20	17	47	47	74	53
21	21	48	–	75	75
22	22	49	45	76	76
23	64	50	50	77	77
24	23	51	51	78	78
25	71	52	52	79	79
26	26	53	73	80	80
27	27	54	54		

Chapter 23: Alternating Current Circuits

New Problems in 6e: 4, 12, 18, 21, 28, 51, 52

5e	6e	5e	6e	5e	6e
1	1	4	3	7	7
2	2	5	37	8	40
3	–	6	6	9	9

Chapter 23 (continued)

5e	6e	5e	6e	5e	6e
10	—	24	44	38	38
11	41	25	25	39	16
12	10	26	{26}	40	8
13	13	27	27	41	11
14	14	28	—	42	20
15	15	29	29	43	33
16	39	30	31	44	—
17	—	31	30	45	34
18	36	32	32	46	46
19	19	33	43	47	47
20	42	34	45	48	48
21	21	35	35	49	49
22	22	36	17	50	50
23	{23}	37	5		

Chapter 24: Electromagnetic Waves

New Problems in 6e:				4, 12, 18, 38, 52, 61, 62	
5e	6e	5e	6e	5e	6e
1	43	21	42	41	7
2	2	22	{22}	42	21
3	—	23	23	43	1
4	3	24	44	44	24
5	5	25	25	45	26
6	6	26	45	46	46
7	41	27	{27}	47	35
8	—	28	28	48	32
9	11	29	51	49	49
10	10	30	30	50	40
11	9	31	{31}	51	29
12	8	32	48	52	53
13	13	33	33	53	—
14	—	34	34	54	54
15	15	35	47	55	55
16	14	36	36	56	56
17	17	37	—	57	57
18	16	38	37	58	58
19	19	39	39	59	59
20	20	40	50	60	60

Chapter 25: The Reflection of Light: Mirrors

New Problems in 6e:				6, 14, 22, 36, 38, 42, 47	
5e	6e	5e	6e	5e	6e
1	1	16	16	31	23
2	32	17	17	32	2
3	3	18	33	33	18
4	4	19	{19}	34	34
5	5	20	20	35	21
6	—	21	35	36	—
7	8	22	—	37	37
8	7	23	31	38	24
9	41	24	—	39	27
10	10	25	{25}	40	28
11	11	26	26	41	9
12	30	27	39	42	43
13	13	28	40	43	46
14	—	29	29	44	44
15	15	30	12	45	45

Chapter 26: The Refraction of Light: Lenses and Optical Instruments

New Problems in 6e:				2, 12, 20, 32, 40, 48, 62, 72, 80, 86, 90, 124, 125	
5e	6e	5e	6e	5e	6e
1	1	27	—	54	114
2	—	28	27	55	117
3	3	29	30	56	56
4	6	30	29	57	57
5	5	31	35	58	116
6	4	32	34	59	61
7	{7}	33	—	60	60
8	8	34	33	61	59
9	105	35	31	62	63
10	108	36	36	63	—
11	13	37	37	64	64
12	11	38	—	65	65
13	{28}	39	39	66	66
14	15	40	38	67	{67}
15	17	41	41	68	—
16	14	42	42	69	101
17	19	43	43	70	70
18	18	44	46	71	71
19	16	46	97	72	68
20	22	47	104	73	113
21	21	48	103	74	117
22	—	49	—	75	75
23	23	50	49	76	106
24	24	51	50	77	77
25	99	52	{51}	78	78
26	26	53	55	79	79

Chapter 26 (continued)

5e	6e	5e	6e	5e	6e
80	–	95	94	110	110
81	81	96	96	111	53
82	82	97	45	112	112
83	98	98	83	113	73
84	–	99	{25}	114	54
85	85	100	100	115	115
86	84	101	69	116	58
87	87	102	102	117	74
88	88	103	47	118	118
89	89	104	44	119	119
90	92	105	9	120	120
91	91	106	76	121	121
92	–	107	107	122	122
93	93	108	10	123	123
94	95	109	109		

Chapter 27: Interference and the Wave Nature of Light

New Problems in 6e: 2, 10, 22, 24, 30, 40, 42, 48, 52, 65, 66

5e	6e	5e	6e	5e	6e
1	47	23	23	45	57
2	46	24	26	46	3
3	–	25	25	47	1
4	4	26	54	48	20
5	5	27	27	49	13
6	6	28	28	50	50
7	{7}	29	29	51	31
8	8	30	–	52	–
9	9	31	51	53	35
10	–	32	33	54	–
11	11	33	32	55	43
12	12	34	34	56	16
13	–	35	53	57	45
14	14	36	36	58	18
15	15	37	37	59	59
16	56	38	39	60	60
17	{17}	39	38	61	61
18	58	40	–	62	62
19	19	41	41	63	63
20	49	42	–	64	64
21	21	43	55		
22	–	44	44		

Chapter 28: Special Relativity

New Problems in *6e*:				8, 18, 28, 30, 48, 49	
5e	*6e*	*5e*	*6e*	*5e*	*6e*
1	3	17	17	33	33
2	38	18	–	34	40
3	1	19	19	35	35
4	4	20	20	36	22
5	{5}	21	21	37	9
6	6	22	36	38	2
7	7	23	27	39	39
8	–	24	24	40	34
9	11	25	25	41	41
10	10	26	26	42	29
11	37	27	23	43	16
12	12	28	42	44	44
13	{13}	29	–	45	45
14	14	30	32	46	46
15	15	31	31	47	47
16	43	32	–		

Chapter 29: Particles and Waves

New Problems in *6e*:				6, 8, 14, 16, 24, 28, 32, 50, 51	
5e	*6e*	*5e*	*6e*	*5e*	*6e*
1	1	18	18	35	35
2	38	19	19	36	36
3	3	20	45	37	21
4	4	21	37	38	2
5	5	22	22	39	23
6	42	23	39	40	–
7	–	24	40	41	–
8	–	25	25	42	7
9	9	26	26	43	27
10	44	27	43	44	10
11	11	28	–	45	20
12	12	29	{29}	46	46
13	13	30	30	47	47
14	41	31	31	48	48
15	15	32	33	49	49
16	–	33	–		
17	{17}	34	34		

Chapter 30: The Nature of the Atom

New Problems in *6e*:				4, 12, 18, 24, 40, 60, 61	
5e	*6e*	*5e*	*6e*	*5e*	*6e*
1	1	21	23	41	41
2	2	22	—	42	42
3	3	23	21	43	43
4	—	24	22	44	29
5	5	25	25	45	7
6	6	26	26	46	34
7	45	27	27	47	47
8	8	28	{30}	48	36
9	9	29	44	49	11
10	10	30	28	50	50
11	49	31	31	51	{51}
12	14	32	32	52	52
13	13	33	33	53	15
14	—	34	46	54	20
15	53	35	{35}	55	55
16	—	36	48	56	56
17	17	37	37	57	57
18	16	38	38	58	58
19	19	39	—	59	59
20	54	40	39		

Chapter 31: Nuclear Physics and Radioactivity

New Problems in *6e*:				2, 30, 38, 44, 54, 61, 62	
5e	*6e*	*5e*	*6e*	*5e*	*6e*
1	1	21	21	41	41
2	—	22	50	42	42
3	3	23	23	43	—
4	48	24	55	44	43
5	5	25	{25}	45	45
6	7	26	26	46	46
7	6	27	27	47	10
8	8	28	28	48	4
9	9	29	29	49	31
10	47	30	—	50	{22}
11	11	31	49	51	18
12	{14}	32	32	52	40
13	13	33	34	53	53
14	12	34	33	54	24
15	15	35	35	55	—
16	16	36	36	56	56
17	17	37	37	57	57
18	51	38	—	58	58
19	19	39	39	59	59
20	{20}	40	52	60	60

Chapter 32: Ionizing Radiation, Nuclear Energy, and Elementary Particles

New Problems in 6e: 2, 10, 22, 28, 29, 38, 46, 54

5e	6e	5e	6e	5e	6e
1	5	19	19	37	—
2	—	20	44	38	36
3	3	21	21	39	39
4	42	22	14	40	40
5	1	23	23	41	4
6	6	24	48	42	41
7	9	25	24	43	43
8	8	26	26	44	20
9	7	27	27	45	31
10	—	28	30	46	—
11	11	29	—	47	47
12	—	30	45	48	25
13	13	31	32	49	49
14	12	32	33	50	50
15	15	33	34	51	51
16	16	34	35	52	52
17	17	35	37	53	53
18	{18}	36	—		

NOTES

NOTES